List of titles

Already published

Cell Differentiation	J.M. Ashworth
Biochemical Genetics	R.A. Woods
Functions of Biological Membranes	M. Davies
Cellular Development	D. Garrod
Brain Biochemistry	H.S. Bachelard
Immunochemistry	M.W. Steward
The Selectivity of Drugs	A. Albert
Biomechanics	R. McN. Alexander
Molecular Virology	T.H. Pennington, D.A. Ritchie
Hormone Action	A. Malkinson
Cellular Recognition	M.F. Greaves
Cytogenetics of Man and other Animals	A. McDermott
RNA biosynthesis	R.H. Burdon
Protein Biosynthesis	A.E. Smith

In preparation

The Cell Cycle	S. Shall
Biological Energy Transduction	C. Jones
Control of Enzyme Activity	P. Cohen
Metabolic Regulation	R. Denton, C.I. Pogson
Polysaccharides	D.A. Rees
Microbial Metabolism	H. Dalton
Bacterial Taxonomy	D. Jones
Molecular Evolution	W. Fitch
A Biochemical Approach to Nutrition	R.A. Freedland
Metal Ions in Biology	P.M. Harrison, R. Hoare
Nitrogen Metabolism in Plants and Microorganisms	A.P. Sims
Cellular Immunology	D. Katz
Muscle	R.M. Simmons
Xenobiotics	D.V. Parke
Plant Cytogenetics	D.M. Moore
Human Genetics	J.H. Edwards
Population Genetics	L.M. Cook
Membrane Biogenesis	J.
Biochemical Systematics	J.
Biochemical Pharmacology	B.
Insect Biochemistry	H

OUTLINE STUDIES IN BIOLOGY

Editor's Foreword

The student of biological science in his final years as an undergraduate and his first years as a graduate is expected to gain some familiarity with current research at the frontiers of his discipline. New research work is published in a perplexing diversity of publications and is inevitably concerned with the minutiae of the subject. The sheer number of research journals and papers also causes confusion and difficulties as assimilation. Review articles usually presuppose a background knowledge of the field and are inevitably rather restricted in scope. There is thus a need for short but authoritative introductions to those areas of modern biological research which are either not dealt with in standard introductory textbooks or are not dealt with in sufficient detail to enable the student to go on from them to read scholarly reviews with profit. This series of books is designed to satify this need. The authors have been asked to produce a brief outline of their subject assuming that their readers will have read and remembered much of a standard introductory textbook of biology. This outline then sets out to provide by building on this basis, the conceptual framework within which modern research work is progressing and aims to give the reader an indication of the problems, both conceptual and practical, which must be overcome if progress is to be maintained. We hope that students will go on to read the more detailed reviews and articles to which reference is made with a greater insight and understanding of how they fit into the overall scheme of modern research effort and may thus be helped to choose where to make their own contribution to this effort. These books are guidebooks, not textbooks. Modern research pays scant regard for the academic divisions into which biological teaching and introductory textbooks must, to a certain extent, be divided. We have thus concentrated in this series on providing guides to those areas which fall between, or which involve, several different academic disciplines. It is here that the gap between the textbook and the research paper is widest and where the need for guidance is greatest. In so doing we hope to have extended or supplemented but not supplanted main texts, and to have given students assistance in seeing how modern biological research is progressing, while at the same time providing a foundation for self help in the achievement of successful examination results.

J.M. Ashworth, Professor of Biology, University of Essex.

RNA Biosynthesis

R. H. Burdon

Department of Biochemistry,
University of Glasgow

LONDON
CHAPMAN AND HALL

A Halsted Press Book
JOHN WILEY & SONS, INC., NEW YORK

First published in 1976
by Chapman and Hall Ltd
11 New Fetter Lane, London EC4P 4EE
© 1976 Alan Smith
Printed in Great Britain by
William Clowes & Sons Ltd,
London, Colchester and Beccles

ISBN 0 412 14050 0

Distributed in the U.S.A.
by Halsted Press, a Division
of John Wiley & Sons, Inc. New York

Library of Congress Cataloging in Publication Data
Burdon, Roy Hunter
RNA biosynthesis

(Outline studies in biology series)
'A Halsted Press book.'
Includes bibliographies and index.
1. Ribonucleic acid synthesis. I. Title.
[DNLM: 1. RNA–Biosynthesis. QU58 B951r]
QP623.B87   574.8'732   75-33182
0–470–12547–0

Contents

1 Introduction

The genetic information for the development and functioning of organisms is encoded in the linear sequences of deoxyribonucleotides that make up its DNA, or deoxyribonucleic acid. It is the *transcription* of such sequence information from this polymer that is the first step in the process of gene expression.

The sequences of deoxyribonucleotides that comprise DNA, whilst physically contiguous, can be envisaged as being operationally divided into discrete regions, or genes. These regions comprise the deoxyribonucleotide sequences required to specify individual macromolecular cell products. The products of DNA *transcription* are also long polymers but made up of ribonucleotides rather than deoxyribonucleotides and are termed ribonucleic acids (RNA). During the *transcription* process the DNA deoxyribonucleotide sequences serve as 'templates' and are reproduced, but in terms of the ribonucleotide sequences of RNA molecules. Certain DNA deoxyribonucleotide sequences, or genes, when transcribed give rise to transfer RNA (tRNA) and others ribosomal RNA (rRNA). The ribonucleotide sequences of these are such as to cause them to fold extensively and, in the case of rRNA to associate with specific proteins to yield the essential cell components called ribosomes.

Transcription of yet other, but special, sequences results in the biosynthesis of messenger RNAs (mRNAs) which contain ribonucleotide sequences which will ultimately be *translated* into new types of sequences, namely, the amino acid sequences of protein molecules (e.g. enzymes) required for cellular function. This switch to a different variety of molecular sequence is complex, but basically each sequence of three ribonucleotides specifies the insertion of one particular amino acid into the polypeptide chain under construction. Whilst mRNA might be considered the means whereby genetic information is actually transmitted from the genome (the DNA) and placed in the appropriate cytoplasmic sites for *translation* into protein, it is important to realise that the actual *translation* machinery (dealt with in detail elsewhere in this series by A.E. Smith in *Protein Biosynthesis*) depends not only on mRNAs for its function but also the presence of other transcription products.

rRNA is necessary as a structural component of the ribosomes upon which translation actually takes place and tRNA is required in amino acid activation, as an adaptor in mRNA-directed amino acid specification and in binding the growing protein chains to the ribosomes. (see Fig. 1.1).

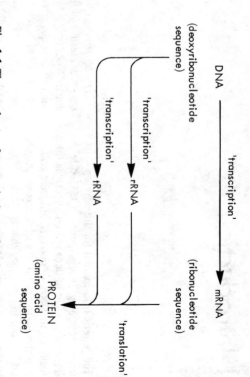

Fig. 1.1 The products of transcription and their cellular role.

2 The products of transcription

2.1 General properties of RNA molecules

An examination of the total RNA isolated from both eukaryotes and prokaryotes shows it to be a collection of polymers of various, but defined, lengths. Chemically they are in fact quite similar to DNA from which they are transcribed. All are long unbranched molecules basically containing four types of nucleotide linked together by 3′–5′ phosphodiester bonds. Unlike DNA however, the nucleotide sugar involved is ribose rather than deoxyribose, and the base uracil usually replaces thymine.

For most RNA molecules the amount of the base adenine (Ade) does not usually equal the amount of uracil (Ura), and the levels of guanine (Gua) and cytosine (Cyt) bases also usually differ from one another. This is quite unlike most DNAs where there is an equivalence of adenine with thymine and guanine with cytosine. This observation was the first indication that most RNA molecules do not possess a regular hydrogen bonded structure like that encountered in DNA, which is composed of complementary polydeoxyribonucleotide sequences arranged in the form of a double helix around a common axis. Such a helical molecule of DNA is maintained in this configuration by the aid of specific hydrogen bonding between the complementary base moieties of the deoxyribonucleotides comprising the two strands, the base adenine (Ade) forming what is known as a complementary base pair with thymine (Thy), and guanine (Gua) forming a complementary base pair with cytosine (Cyt).

With regard to the structures of RNA molecules, or polyribonucleotides, (see Fig. 2.1) it has become usual to use the convenient 'short-hand' system illustrated in Fig. 2.2. A vertical line denotes the carbon chain of the sugar with a base attached at C–1′. A diagonal line from the middle of a vertical line indicates the phosphate link at C–3′ while one at the bottom of a vertical line denotes the phosphate link at C–5′. An even simpler system is also used. A phosphate group is denoted by p. When placed at the right of the nucleoside symbol the phosphate is esterified at the C–3′ of the ribose moiety, whereas when it is placed to the left of the nucleoside symbol, the phosphate is esterified at the C–5′ of the ribose moiety. Thus pUpU is a dinucleotide with one phosphate esterified at the C–5′ of a uridine residue and a phosphodiester bond between C–3′ of that residue and C–5′ of the adjacent residue. Sometimes the letter p is replaced by a hyphen, e.g. pU–U.

Whilst DNA has a firmly established helical structure, the nature of the secondary and tertiary structure of RNA molecules is less well defined. In solutions of low ionic strength, RNA molecules behave like typical highly swollen polyelectrolyte chains, but an increase in ionic strength causes the chains to contract upon themselves so as to display relatively low intrinsic viscosities and high sedimentation rates. This suggests the existence of at least some helical

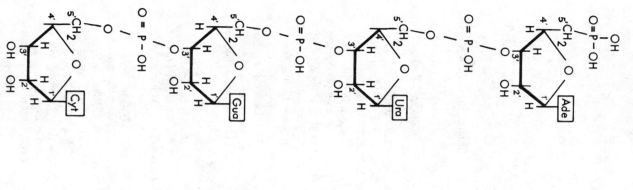

Fig. 2.1 Structure of a hypothetical polyribonucleotide.

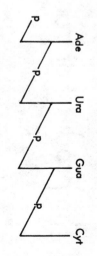

or

pApUpGpC

or

PA–U–G–C

Fig. 2.2 Shorthand notations for hypothetical polyribonucleotide shown in Fig. 2.1. Nucleoside abbreviations; A, adenosine; G, guanosine; U, uridine; C, cytidine.

regions with complementary base pairing in certain regions of RNA nucleotide chains. Enzymological techniques now permit the formation in the test-tube of RNA molecules in which the only base is adenine. These are referred to as poly (A) (polyadenylic acid). RNAs whose only base is uracil can also be synthesised, namely poly (U) (polyuridylic acid). When equimolar amounts of these somewhat artificial RNAs are mixed in solution they form a complex known as poly (A) · poly (U) (see Fig. 2.3 in which the adenine bases of one strand are linked by hydrogen bonds to the complementary uracil bases of the other strand [1]. Indeed the X-ray diffraction pattern of this complex indicates a double helical structure as in DNA with 10 bases per turn of the helix, the pitch of which is 3.4 nm.

In fact this RNA helix behaves like a DNA helix in many ways. For instance it shows the phenomenon of 'molecular melting' or 'helix-coil' transition. When heated in 0.15M NaCl at neutral pH the absorbance at 260 nm rises sharply at a temperature of around 60°, the so-called melting temperature, or T_m. At the same temperature the specific optical rotation at 589 nm decreases rapidly. These effects are due to the separation of the two

membranes and membrane bound organelles present in the eukaryotic cell. Another basic difference is that in prokaryotic cells there is no nucleus. The chromosome comprises a single very long DNA molecule (often cyclic) not clearly associated with any other cell component other than the membrane at one region. In eukaryotic cells the DNA is first enclosed within the nucleus and furthermore it is there associated with considerable amounts of basic (histone) and other proteins to form the chromosomes which condense and become clearly visible at the time of mitosis. However during interphase (as in Fig. 2.4) there is much less structural detail to be seen either by light or electron microscope. Nevertheless two different states have been recognised namely *euchromatin*, which consists of loosely coiled fibres of DNA and protein, and *hetero-chromatin*, which appears to comprise tightly coiled fibres [4].

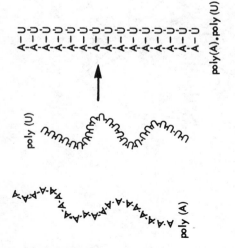

poly(A).poly(U)

Fig. 2.3 The association of single strands of poly (A) and poly (U) to form the helical poly (A) · poly (U) complex in which two strands are linked by hydrogen bonding between A and U.

complementary strands of the helix on heating. Cooling reverses these effects and the helix is reformed [1].

When solutions of naturally occurring RNAs are heated quite similar, but less pronounced changes occur, which suggest that the RNA chains are folded back upon themselves in a number of places to form short helical regions [2]. Our understanding of these structural considerations is becoming clearer now that several distinct species of cellular RNA have been isolated and extensively characterised.

2.2 The intracellular locations of RNA

As already mentioned there is a variety of cellular RNA species and these often have quite discrete intracellular locations. The basic cytological features of typical prokaryotic and eukaryotic cells are summarised in Figs. 2.4 and 2.5. (A detailed summary of modern cytological work can be obtained in [3]).

One difference between prokaryotic and eukaryotic cells is the complex systems of

While the composition of cellular components can be studied *in situ* with the use of various sophisticated histochemical and cytochemical techniques, considerable information on cellular processes such as RNA biosynthesis has been obtained by study of cell components *after* disruption of the cells followed by fractionation involving differential centrifugation. There is no standard method however that is applicable to all cells. Nevertheless the methodology that has been developed for rat liver cells [5] has proven remarkably versatile and adaptable to other tissues and organisms [6].

Basically the tissues (or cells) are firstly disrupted mechanically in suitable media (usually containing sucrose to reduce aggregation problems). This can be done with a tissue grinder, glass homogeniser, a Waring blender or other high speed mixer with the aid of abrasives or glass beads. The disrupted cell preparation can then be centrifuged at around 200 g to remove nuclei and general cell debris

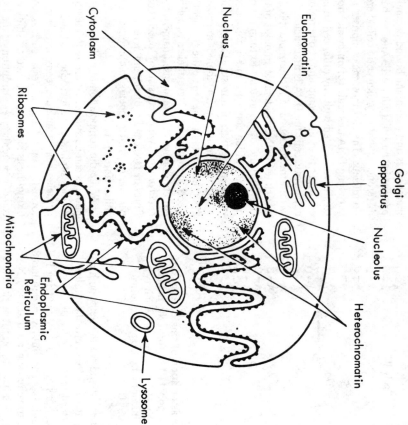

Fig. 2.4 A highly schematic representation of a eukaryotic (e.g. mammalian) cell.

Cytoplasm

Nucleus

Euchromatin

Golgi apparatus

Nucleolus

Heterochromatin

Ribosomes

Mitochondria

Endoplasmic Reticulum

Lysosome

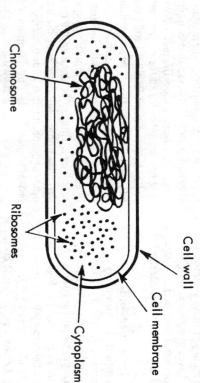

Fig. 2.5 A highly schematic representation of a prokaryotic (e.g. bacterial) cell.

Chromosome

Ribosomes

Cell wall

Cell membrane

Cytoplasm

including unbroken cells. After removal of the nuclear fraction the remaining material is centrifuged at 8500 g for 10 min or so to bring down mitochondria, and at 140 000 g for 90 min to sediment microsomes and any 'free' (or 'unbound') ribosomes. The clear supernatant fluid is said to correspond to the cell sap (or cytosol).

Microsomes, (which correspond to fragments of endoplasmic reticulum) when treated with the detergent, sodium deoxycholate, are further disrupted into a non-sedimentable portion derived from the membranous component which contains most of the protein and phospholipid, and a particulate portion sedimentable at 14 000 g which corresponds to the

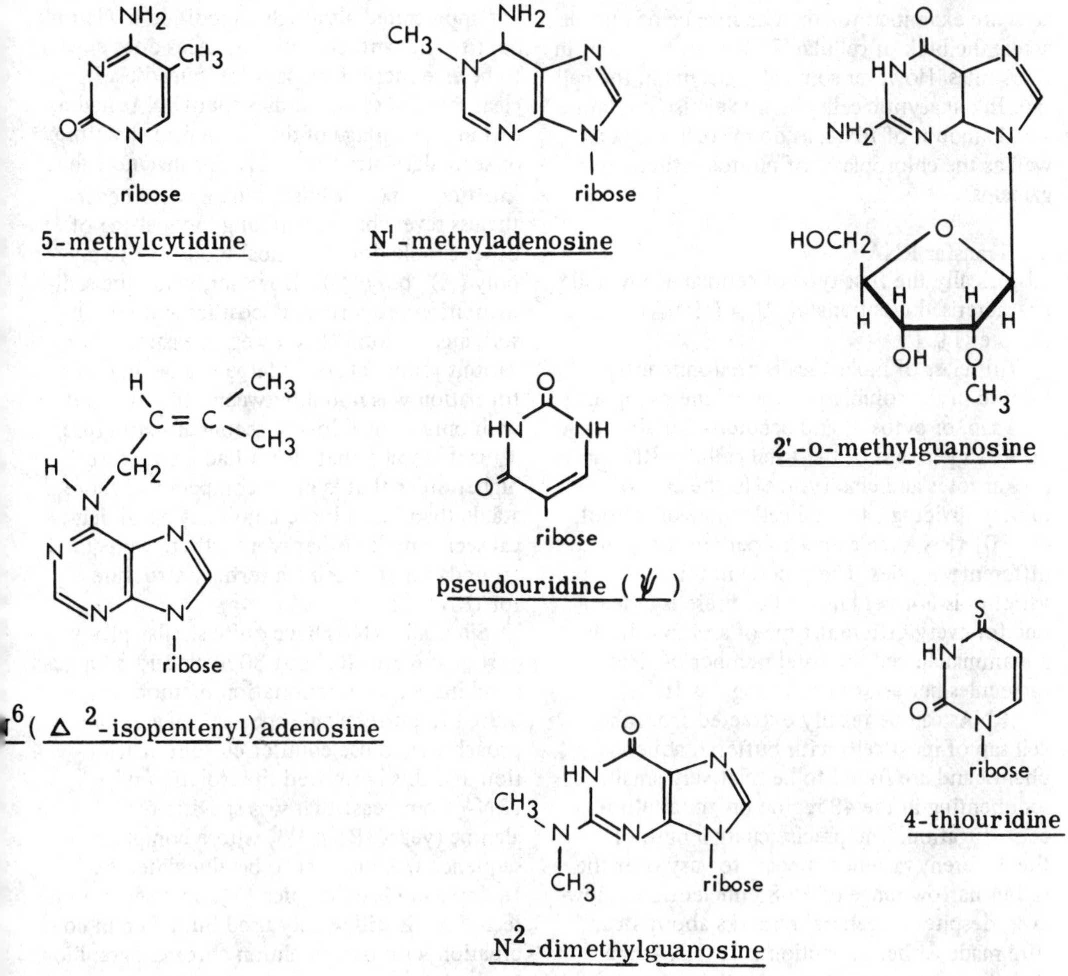

2'-o-methylguanosine

4-thiouridine

N^1-methyladenosine

pseudouridine (ψ)

N^2-dimethylguanosine

5-methylcytidine

^{6}(Δ^2-isopentenyl) adenosine

Fig. 2.6 The structure of a few modified nucleosides.

'bound' ribosomes (i.e. the ribosomes original-ly bound as part of the endoplasmic reticulum as distinct from those 'free' in the cytoplasm).

As stated already the basic procedure of differential ultracentrifugation has many varia-tions and for further details of the technique the reader is referred elsewhere [6].

When the above mentioned subcellular frac-tions from both prokaryotic and eukaryotic cells are examined for the occurrence of nucleic acids, the bulk of cellular RNA is to be found in ribosomes. However some also occurs in the cell sap. In eukaryotic cells the nuclei also contain a small amount of RNA, as do mitochondria as well as the chloroplasts of photosynthetic or-ganisms.

2.3 Transfer RNA

Historically the first type of cellular RNA to be characterised was transfer RNA (tRNA) (see Chapter 1).

This class of molecules is predominantly located in the soluble portion of the cytoplasm (cell sap, or cytosol) and accounts for about 10 to 15 per cent of the total cellular RNA in prokaryotes and eukaryotes. In the case of rapidly dividing bacterial cells there are about 4×10^5 tRNA molecules of perhaps fifty or so different varieties. The precise number of varieties is not yet known but there is at least one for every different type of amino acid. In a mammalian cell the total number of tRNA molecules per cell can be as high as 10^8.

tRNAs can be readily extracted from the cell sap of most cells with buffered aqueous phenol and are found to be relatively small, sedimenting in the 4S region on zonal ultra-centrifugation. The precise chain length of the different varieties appears to vary over the rather narrow range of 76-85 nucleotides. How-ever, despite the general remarks about struc-ture made earlier in Section 2.1, tRNAs are slightly exceptional. Although they contain the four common nucleosides they contain also a variety of unusual or 'minor' ribonucleosides (see Fig. 2.6). Whilst the presence of these 'minor' nucleosides may at first sight be both puzzling and confusing, it will be seen later in this book that they arise as simple structural modifications to the primary nucleotide se-quence of newly made RNA molecules. Despite this chemical knowledge, the functional signi-ficance of these molecular modifications is not yet appreciated although a modified nucleoside next to the 'anti-codon' sequences does appear to be essential for recognition purposes. It was clear from physical studies that tRNAs had a certain percentage of double helical structure or secondary structure [2]. For instance, the 'melting' curves, whilst gradual, were never-theless reversible, resembling the melting of other simple double helical molecules (e.g. poly (A) · poly (U)). However, when the sedi-mentation properties, viscosities and UV abs-sorbance of total tRNA were measured at various temperatures, a large change in con-formation was noted between $20°-40°$, but with only a small loss in secondary structure. Thus it seemed that tRNA had a structure of higher order that is more compact and more stable than just a loose combination of heli-cal segments. In other words, there were good grounds for supposing a tertiary structure for tRNA [2].

Since all tRNAs have quite similar pro-perties and are all about 80 nucleotides long a combination of fractionation methods is re-quired to purify single species. An early ap-proach was to use counter current distribu-tion and this permitted the isolation of tRNA from yeast that was specific for alanine (yeast tRNA^{Ala}) whose complete sequence was the first to be elucidated by Holley and his colleagues [7]. This separation technique is still widely used but often in com-bination with other column chromatographic methods.

While the original sequence determination

required fairly large amounts of pure tRNA, only 0.5 mg of highly labelled [^{32}P]–tRNA is required for the extremely rapid sequencing method developed by Sanger and his colleagues – the so-called 'fingerprinting' technique [8]; The methods employed in primary sequence determinations consist essentially of the controlled degradation of the RNA with enzymes and separation of the products by chromatography [7], or by two-dimensional electrophoresis in the case of the Sanger technique [8].

The determination in 1965 of the complete sequence of yeast tRNA^{Ala} has been followed by the sequence of a further forty tRNAs [9]. Just about all these sequences can be fitted to the same hydrogen bonded secondary structure of loops and short helical regions as shown Fig. 2.7. The two parts of the structure which have a known function are (a) the 3'–terminal adenosine residue to which the amino acid is esterified, and (b), the three adjacent bases carrying the *anticodon* which in all tRNAs occupy the same position in the clover leaf.

Whilst there is a lot of evidence to support the clover-leaf arrangement in two dimensions, only recently has there been any unambiguous data regarding tertiary structure. The results of a systematic crystallisation study of yeast tRNA^{Phe} were crystals suitable for X-ray analysis at 3Å resolution using the method of isomorphous replacement [10]. A schematic diagram is shown in Fig. 2.8 of the basic three dimensional arrangement found.

2.4 The RNAs of the ribosomes

Whilst a great deal is known about tRNA, its structure and function, the bulk of cellular RNA (about 80 per cent) is contained in the minute cytoplasmic particles known as ribosomes. These have a diameter of around 20 nm, contain protein as well as RNA, and are found in all types of living cell both 'free' and 'bound' to

membrane components (e.g. endoplasmic reticulum).

It is customary, although perhaps somewhat prosaic, to characterise the ribosomes by their sedimentation coefficients expressed in Svedberg units. In mammalian cells there are around 5×10^6 ribosomes each sedimenting around 80S; however in bacteria the basic ribosome is only 70S and there are only $15-18 \times 10^3$ per cell.

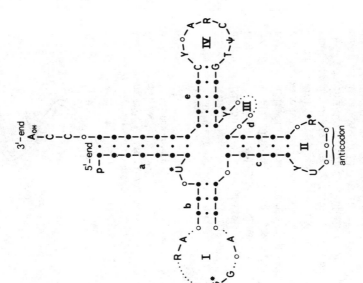

Fig. 2.7 A generalised version of the clover leaf model for tRNA showing constant features. ●, any base; R, purine; Y, pyrimidine; T, ribothymidine; ψ, pseudouridine; R*, modified adenine.

Ribosomes from all sources have very similar structures [11]. They are roughly ellipsoidal complexes of RNA and proteins (ribonucleo-

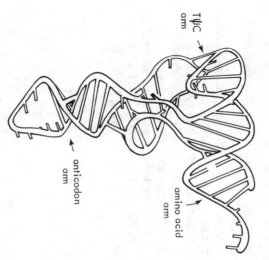

TψC arm

anticodon arm

amino acid arm

Fig. 2.8 A schematic model of yeast phenyl-alanine transfer RNA (the ribose phosphate backbone is drawn as a continous cylinder with bars to indicate hydrogen bonded base pairs).

proteins) of molecular weight ranging from 2.5×10^6 in bacteria to roughly 4×10^6 in mammals. Basically they comprise two sub-units of which the larger is 2 to 2.5 times the size of the smaller. The two subunits associate together and function as an integrated unit in protein synthesis. They dissociate between rounds of protein synthesis *in vivo* and can be made to dissociate *in vitro* by exposure to low Mg^{2+} concentrations, high concentrations of mono-valent cations, or EDTA. The relative S values for the ribosome and its subunits (large and small) are: in bacteria 70S, (50S and 30S) and in higher organisms 80S, (60S and 40S) (See Fig. 2.9).

Somewhat more than half the mass of each ribosomal subunit consists of RNA, the re-mainder being made up of protein. The small subunit comprises a single RNA molecule (0.55×10^6 daltons — or 16S—in bacteria and

0.75×10^6 daltons — or 18S — in mammalian ribosomes) together with several proteins (20 in *E.coli*). The large subunit contains one large RNA molecule (1.1×10^6 daltons — or 23S — in bacteria, and 1.75×10^6 daltons — or 28S — in mammalian ribosomes) as well as one small RNA molecule called the 5S RNA (120 nucleotides long) and a further group of proteins (36 in *E.coli*). Fractionation of the pro-tein from animal cell ribosomes also indic-ates a high degree of complexity [11].

Bacteria, actinomycetes, blue green algae and higher plant chloroplasts however all have ribosomal RNAs of molecular weight 1.1×10^6, whereas the corresponding values for higher plants, ferns, algae, fungi and some protozoa are 1.3×10^6 and 0.7×10^6 [14]. The 0.7×10^6 component (18S) is common to all animals, but the large (28S) component has evolved with each major step of animal evolution from 1.4×10^6 in sea urchins to 1.75×10^6 in mammals [14]. Whilst the ri-bosomal RNA from most organisms possess broadly similar base compositions with guanine plus cytosine contents of between 50–60 per cent there are some exceptions to this e.g. *Drosophila* (40 per cent), *Tetrahymena* (43 per cent) [11].

As was the case for tRNA a small number of specific nucleotides of ribosomal RNAs are modified following transcription. *E.coli* 16S and 23S RNA possess 22 and 27 methyl groups respectively. Most of these groups are on various base moieties (see Fig. 2.10). Whilst the large RNAs (28S and 18S) of the mammalian ribosome are also methylated (71 and 46 methyl groups respectively), unlike the bac-terial situation, most (95 per cent) are on ribose moieties and the rest on bases [11].

The question of whether or not the poly-nucleotide chains of ribosomal RNA (rRNA) are covalently continuous has been extensively examined. The sedimentation properties of

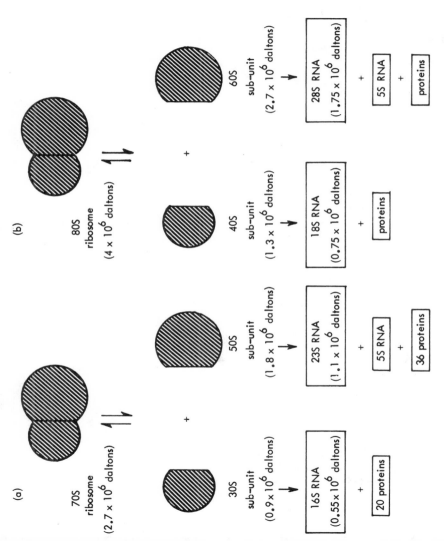

Fig. 2.9 S-values and molecular weights of the components of (a) a ribosome from *E.coli* and (b) a mammalian ribosome.

carefully purified *E.coli* 16S and 23S RNAs remain unchanged after heating or chemical denaturation. The same is true of mammalian rRNAs with one qualification; on treating 28S RNA by brief exposure to high temperature or with reagents which disrupt hydrogen bonds a small 5.8S fragment of RNA is released [28]. This fragment appears to be hydrogen bonded to the parent 28S rRNA molecule and its origin will be discussed in Chapter 3. Such findings suggest that the polynucleotide chains of rRNA

are covalently continuous, with the exception of the 5.8S fragment of 28S RNA. This is supported by direct length measurements of rRNA molecules by electron microscopy after denaturation in 8M urea [29].

At present there is only indirect evidence relating to the three dimensional arrangement of ribosomal RNAs and proteins in the ribosome. Physico-chemical data suggest the RNA contains alternating single and double helical regions (see Fig. 2.10), and that much of the ribosome surface

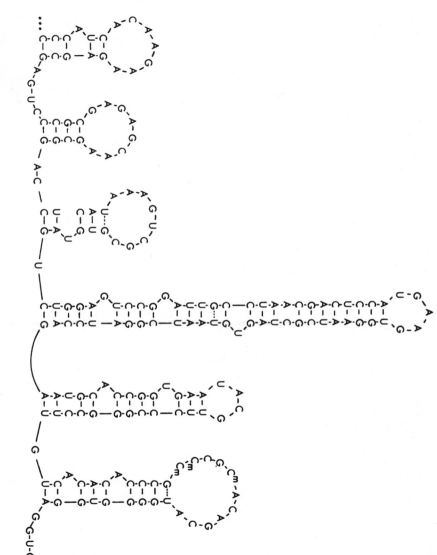

Fig. 2.10 An example of the possible secondary structure of a small portion of the 16S *E.coli* ribosomal RNA [27]. (a small 'm' denotes a nucleoside residue which has been methylated).

is in fact RNA presumably facilitating interaction of rRNA with other RNA components of the protein synthesis system [13].

Having considered the larger rRNAs, note should be made of the small 5S ribosomal RNA which occurs on all large ribosomal subunits (except possibly those in mitochondria, Section 2.6). This small RNA species should not be confused with the 5.8S RNA fragment mentioned earlier which is hydrogen bonded to 28S rRNA. 5S RNA is bound, but not covalently, to some other ribosomal com-

ponent (perhaps a protein). It contains no modifi[ed] nucleotides and a proposal is that it may be a stru[ct]ural part of the 'A'-site for binding amino-acyl tRNA.

E.coli 5S rRNA (see Fig. 2.11) was in fact the first 32P labelled RNA to be sequenced by Sanger and his colleagues [15]. From an evolutionary point of view it seems to have been rigorously conserved since both prokaryotes and eukaryotes have 5S RNA species 120 nucleotides long. On the other hand there are some small variations

geneous in size. In *E.coli* [17] for instance there are about 1000 mRNA molecules, and whilst the average protein chain is between 300–500 amino acids in length, the average size of the corresponding mRNAs would thus be 900–1500 nucleotides. However, some mRNAs are known to carry information for more than one protein. These are termed *polycistronic* and clearly contain more nucleotides. In addition there is evidence for additional non-coding sequences in mRNAs whose function is not known at present. Most mRNAs from mammalian cells also contain non-coding sequences but are probably monocistronic (i.e. only specify one protein) [18].

Whilst the original experiments suggested mRNAs to have a short half-life (1–2 min), the metabolic stability of mRNA is now known to vary quite widely from cell to cell (in HeLa cells it is 7 to 24 hr [20]).

With the exception of certain viral RNAs it had been quite difficult to obtain homogeneous preparations of natural mRNAs which would direct the synthesis of specific proteins in cell-free *in vitro* systems. However by the use of newer techniques such as purification of RNA by acrylamide gel electrophoresis coupled with judicious choice of biological material it was soon possible to obtain pure or partially pure preparations of mRNAs for globin (9S), histone (9–10S), lens crystallin (17S), immunoglobulin (12–14S), ovalbumin (15S), myosin (26S) to mention a few [20].

One of the most conspicuous structural features of some eukaryotic mRNAs is a long uninterrupted sequence of adenosine nucleotides attached at their 3'-end. This polyA tail has been reported to vary from 60 up to 200 nucleotides depending on the mRNA [20]. Whilst mRNA from lower eukaryotes such as slime moulds and yeast also have polyA tails, the histone mRNA of higher eukaryotes is specifically known to lack such a tail [20].

Whilst the function of the polyA remains

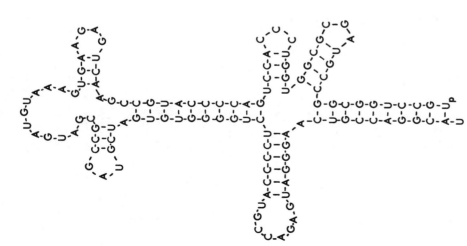

Fig. 2.11 A possible secondary structure for *E. coli* 5S ribosomal RNA.

in base composition (the guanine plus cytosine content varies between 52–64 per cent) and sequence [16].

2.5 The message

Messenger RNA (mRNA) is now known to account for 3–5 per cent of the cellular RNA and is characterised by its reversible association with ribosomes forming polysomes. It is hetero-

obscure it has proven useful for isolating mRNAs. For example RNAs with polyA tails can be bound tightly to columns of Sepharose to which polyU has been attached and can thus be separated from other contaminating RNAs [20].

As the total mRNA of a cell comes from many different genes and encodes protein of a wide variety of sequences, it might be expected to have a base composition resembling that of the cell's DNA. Whilst observation has indicated this to be broadly correct and serves to distinguish the total cell mRNA population from rRNA or tRNA, it is certainly not true for individual messengers like globin or lens crystallin mRNAs. Indeed from the amino acid composition of silkworm fibroin the somewhat unusual base composition of silkworm fibroin mRNA was predicted and confirmed upon its subsequent isolation from the silk gland of *Bombyx mori* (40 per cent guanine and 19 per cent cytosine) [21].

Very little is known about the secondary structure of mRNA. Globin mRNA shows a hyperchromicity upon heating. Sedimentation and electrophoretic studies together have indicated myosin and ovalbumin mRNAs to have more extended chains than is the case for ribosomal RNA. There are however suggestions that globin mRNA may in fact occur in a variety of configurations [10].

2.6 The RNAs of eukaryotic organelles

2.6.1 *Mitochondrial RNA*
Not only do mitochondria contain their own DNA generally in the form of double helical cyclic molecules of about $5\,\mu$ (that from fungi is about five times that size), they also contain their own ribosomes, tRNA and apparently mRNAs [22,23].

Mammalian mitochondria contain ribosomes that sediment at 55S (fungal ribosomes are bigger, 73S). The 55S ribosome dissociates into 40S and 30S subunits and contains RNAs of 0.56×10^6 daltons (21S) and 0.36×10^6 daltons (12S) respectively and which contain modified nucleosides. A notable difference however between mitochondrial ribosomes and cytoplasmic ribosomes is their apparent lack of a small 5S rRNA [22].

Mitochondria contain tRNAs that differ from the corresponding tRNAs of the cell sap although they are still of the same size range and contain modified nucleotides. A notable difference is the presence of $tRNA_f^{Met}$, which is absent from the cell sap [22].

With regard to the occurrence of mRNA species in mitochondria there are data indicating a number of discrete species terminating in 80 nucleotide long polyA tails in mitochondria of mammals and insects [24] (yeast mitochondrial mRNA may not have polyA tails [25]).

2.6.2 *Chloroplast RNAs*
Like mitochondria, chloroplasts also contain cyclic double helical DNA molecules ($40\,\mu$), as well as ribosomes and some specific tRNAs [26]. The two large ribosomal RNAs from ribosomes of higher plant chloroplasts sediment at around 70S are 1.1×10^6 daltons respectively [14]. Additionally there is a detectable 5S rRNA associated with the large subunit. Although plant mRNA is known to have polyA [19] there is no clear indication yet whether the mRNA reported to exist in chloroplasts contains polyA or not.

References

[1] Steiner, R.F. and Beers, R.F. (1961) *Polynucleotides*, Elsevier Amsterdam.

[2] Cramer, F. (1971), *Prog. Nucleic Acid Res. Mol. Biol.*, **11**, 391.

[3] Novikoff, A.B. and Holtzman, E. (1971), *Cells and Organelles*, Holt, Rinehart and Winston, New York.

[4] Ord, M.G., and Stocken, L.A. (1973), in *Cell Biology in Medicine*, Ed. E.E. Bittar,

p. 151, John Wiley and Sons, New York

[5] Hogeboom, G.H. and Schneider, W.C. (1955), in *The Nucleic Acids*, Eds. Chargaff, E. and Davidson, J.N., p. 199, Academic Press, New York.

[6] Murray, R.K., Suss, R. and Pitot, H.C. (1967), in *Methods in Cancer Research*, **2**, Ed. Busch, H., p. 239, Academic Press, New York.

[7] Holley, R.W., Apgar, J., Everett, G.A., Madison, J.T., Marquisee, M., Merrill, S.H., Penswick, J.R. and Zamir, A. (1965), *Science*, **147**,1462.

[8] Sanger, F., Brownlee, G.G. and Barrell, B.G. (1965) *J. Mol. Biol.*, **13**,373.

[9] Barrell, B.G. and Clark, B.F.C. (1974), *Handbook of Nucleic Acid Sequences*, Joynson-Bruvvers, Oxford.

[10] Robertus, J.D., Ladner, J.E., Finch, J.T., Rhodes, D., Brown, R.S., Clark, B.F.C. and Klug, A. (1974), *Nature*, **250**, 546.

[11] Maden, B.E.H. (1971), in *Prog. Biophys. Mol. Biol*, **22**,127.

[12] Kurland, C.G. (1972), *Ann. Rev. Biochem.*, **41**,377.

[13] Cox, R.A. (1966), *Biochem. J.*, **98**, 841.

[14] Loening, U.E. (1968), *J. Mol. Biol.*, **38**, 355.

[15] Brownlee, G.G., Sanger, F. and Barrell, B.B. (1967), *Nature*, **215**,735.

[16] Ford, P.J. (1973), *Biochem. Soc. Symp.*, **37**, 69-81.

[17] Singer, M.F., and Leder, P. (1966), *Ann. Rev. Biochem*, **35**,195.

[18] Matthews, M.B. (1973), *Essays in Biochem.*, Vol. 9, Academic Press, London and New York.

[19] Sagher, D., Edelman, M., and Jakob, K.M. (1974), *Biochim. Biophys. Acta*, **349**, 32.

[20] Brawerman, G. (1974), *Ann. Rev. Biochem.*, **43**,621.

[21] Suzuki, Y. and Brown, D.D. (1972), *J. Mol. Biol.*, **63**, 409.

[22] Borst, P. (1972), *Ann. Rev. Biochem.*, **41**, 333.

[23] Kroon, A.M. and Saccone, C. (1974), *The Biogenesis of Mitochondria*, Academic Press, New York.

[24] Hirsch, A., Spradling, A. and Penman, S., (1974), *Cell*, **1**,31.

[25] Groot, G.S.P., Flavell, R.A., Van Ommen, G.J.B., Grivell, L.A. (1974), *Nature*, **252**, 167.

[26] Sager, R. (1972), *Cytoplasmic Genes and Organelles*, Academic Press, New York.

[27] Ehresmann, C., Stiegler, P., Mackie, G.A., Zimmermann, R.A., Ebel, J.P. and Fellner, P. (1975), *Nucleic Acids Research*, **2**, 265.

[28] Pene, J.J., Knight, E. and Darnell, J.E. (1968), *J. Mol. Biol*, **33**, 609.

[29] Granboulan, N. and Scherrer, K. (1969), *Europ. J. Biochem.*, **9**, 1.

3 Patterns of RNA biosynthesis in the cell

A noticeable feature of eukaryotic organisation to emerge from the previous chapter is that although the DNA (with the exception of mitochondrial DNA) is confined to the nucleus, the bulk of the cellular RNA is located in the cytoplasm. Thus the question arises as to the role of the nucleus in RNA biosynthesis. Autoradiographic experiments involving the use of radioactively labelled ribonucleosides were designed to follow RNA synthesis [1]. This approach coupled with cell fractionation involving differential centrifugation of homogenates of cells variously exposed to the labelled ribonucleosides, indicated the nucleus and not the cytoplasm to be the site of synthesis. Moreover, studies at the enzymic level indicated the presence of enzyme activity exclusive to the nucleus which was not only capable of synthesising new RNA from ribonucleoside 5'-triphosphate units, but which also apparently took its directions for this synthesis from the nuclear DNA which acted as 'template' [2,3]. As will be detailed in Chapter 4, it was soon shown that the products of the nuclear RNA synthesising enzymes (now known as DNA-dependent RNA polymerases) have ribonucleotide sequences complementary to *one* of the strands of the DNA which was used as 'template'. A guanine residue in the DNA template strand dictates the insertion of a cytosine nucleotide in the RNA strand under construction whilst a cytosine in the DNA causes a guanine nucleotide to appear in the new RNA strand. Similarly a thymine in the DNA results in an adenine nucleotide in the RNA and an adenine in the DNA, a uracil nucleotide in the RNA (see Chapter 4 for further details).

3.1 The concept of post-transcriptional processing

Whilst most of the cellular RNA of eukaryotes is cytoplasmic, the nucleus does contain some RNA (about 5 per cent of the total) as mentioned previously (2.2). It turns out, however, that its existence is mainly ephemeral, and there is now substantial evidence indicating it to include several species which are intermediates in biosynthetic pathways leading to the formation of tRNAs, ribosomal RNAs and messenger RNAs of the cytoplasm. The primary structure of these intermediates is modified in certain instances by cellular enzyme systems subsequent to the completion of their transcription from the DNA template.

Such modifications include (a), 'trimming' or 'tailoring', that is alteration to the length of the primary transcription product by scission mechanisms, or (b), the alteration of primary nucleotide sequences as a result of base or sugar modification, or (c), the addition of specific nucleotide sequences. This chemical editorial work carried out by the cell is a prerequisite in the formation of certain nucleic acid species both in eukaryotic as well as prokaryotic cells and the molecular events involved can be collectively described as *post-transcriptional processing events* [4].

3.2 Ribosomal RNA biosynthesis

3.2.1 The ribosomal RNA precursor

Further support for the concept that the composition and nucleotide sequence of cytoplasmic RNA was under the control of nuclear DNA was obtained from molecular hybridisation experiments in which it was shown that there were sequences of DNA precisely complementary to the nucleotide sequences of ribosomal RNA, tRNA and messenger RNA. The technique of molecular hybridisation — in which the extent of *in vitro* formation of stable hydrogen bonded complexes between purified RNA molecules and complementary regions on isolated DNAs can be determined — after denaturation to single strands has been of great value. The principles and practice involved have been well reviewed [5,6] (see also Fig. 3.10).

In bacteria, the deoxyribonucleotide sequences complementary to ribosomal RNA are found to amount to approximately 0.1—0.2 per cent of the total DNA genome [7]. Thus the bacterial genome possesses roughly ten copies of the genes for ribosomal RNA. By similar molecular hybridisation tests, animal cells possess DNA complementary to several hundred copies of ribosomal RNA. (e.g. 260 in *Drosophila*, 900 in *Xenopus* 1100 in HeLa cells [8]).

Whilst there is RNA in the nucleus of eukaryotes, most of this is located in the nucleolus (about 3 per cent of the total cellular RNA). In fact the nucleolus is very active in the manufacture of RNA [1,9]. Moreover, molecular hybridisation experiments with certain eukaryotes demonstrate that the parts of the DNA genome which contain the deoxyribonucleotide sequences complementary to the sequences that make up ribosomal RNA are located in the nucleolus, or nucleolar organiser region of the genome [10]. Evidence from further hybridation tests implicate the nucleolus in ribosome formation. Wild type *Xenopus*

laevis possess two nucleoli, one per haploid set of chromosomes. A mutant is known which possesses only one nucleolus and this can be used to breed anucleolate embryos. Such embryos are unable to produce ribosomes, but receive sufficient maternal ribosomes from the oocyte to support early development only. The mutants lack the nucleolar organiser together with all the DNA which is complementary to ribosomal RNA (sometimes called ribosomal DNA or rDNA). The mutants carrying one nucleolus possess exactly half the wild type complement of ribosomal DNA [11].

There is now evidence that the multiple genes for ribosomal RNA are arranged as linear arrays of contiguous units. Deletions of the nucleolar organiser, which comprises a very small segment of a chromosome, results in the complete loss of a haploid content of ribosomal DNA, i.e. a few hundred genetic units.

Additionally, ribosomal DNA has been shown to contain equimolar amounts of sequences complementary to 28S and 18S ribosomal RNA and in *Xenopus* and *Drosophila* at least, it can be shown that the sequences for 28S and 18S alternate within the ribosomal DNA [12]. No evidence for such intermingling is available at the DNA level for a mammalian system, but it seems likely that such a basic feature would be conserved during evolution.

Our understanding of the molecular events occurring in the nucleolus that give rise to ribosomal RNA owes much to the intensive efforts of Darnell and Penman and their respective colleagues during the last few years [9]. In 1962 Scherrer and Darnell (see [9]) demonstrated the existence of relatively short lived species of RNA in cultures of human cells exposed for very short time to ^{3}H-labelled uridine. This RNA occurred in only small amounts and was distinguishable from tRNA and ribosomal RNA. It sedimented at 45S and its molecular weight is now known to be about

4.1 × 10⁶. Although the site of its synthesis was soon found by cell fractionation techniques to be the nucleus [13] a more interesting feature was discovered. Shortly after its synthesis, or transcription from the nucleolar DNA, this long, single-stranded molecule undergoes a series of molecular 'tailoring' events in the nucleolus whereby its molecular dimensions are progressively reduced to give rise eventually to the 28S and 18S ribosomal RNA species found in cytoplasmic ribosomes.

Whilst the 45S is found exclusively in the nucleolus, after longer labelling times radio-activity begins to appear in other species of RNA. Firstly in 41S and 32S RNA, which are also confined to the nucleolus, then 20S RNA, and then in 18S RNA which passes rapidly to the cytoplasm. Finally radioactivity appears in 28S RNA initially in the nucleolus, then the nucleoplasm and shortly afterwards in the cytoplasm. A precursor-product relationship (see

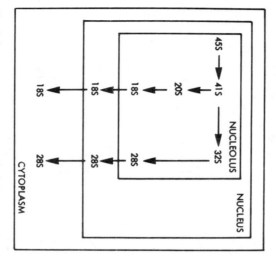

Fig. 3.1 A schematic diagram of the intracellular location of the basic steps believed to be involved in the post-transcriptional processing of the 45S ribosomal precursor in HeLa cells.

Fig. 3.1) between these RNAs was proposed [9] as a result of the use of actinomycin D. This antibiotic at low concentration of 0.05 µg/ml of culture blocks the synthesis of the 45S species in HeLa cells thereby achieving a means of assessing the cellular fate of 45S RNA without the complication of further synthesis. Radio-activity appearing initially in the 45S component is lost on actinomycin addition but re-appears almost simultaneously in both the 32S and 20S regions. Subsequently that in the 32S region shifts to a position corresponding to the 28S RNA, and that in 20S to the 18S RNA.

Recently our understanding of topology of the 45S molecule has been greatly clarified by virtue of some sophisticated electron micro-scopy. Previously our ideas regarding the rela-tive arrangements of the 18S, 28S and non-ribosomal sequences derived from various end-group analyses. However, a new technique of secondary structure mapping has now been developed by Wellauer and Dawid [14] to order these sequences. When the precursor is spread for electron microscopy 'hairpin' loops may form between short complementary sequences and the position of each loop along the linear molecule provides a map of it. The 28S rRNA has a characteristic map and is present at the 3'-end of the precursor and 18S RNA is characterised by an absence of loops. Using this approach the topology of the 45S precursor is as shown in Fig. 3.2. Also indicated in Fig. 3.2 are the sites of cleavage occurring in nucleoli that generate the intermediary species and final-ly the mature ribosomal RNAs. With regard to the sequence modifications to rRNA (i.e. the methylations and pseudouridine formation) the majority of these occur rapidly in the nucleolus on the 45S precursor. Interestingly the methy-lations occur only in the portions of the pre-cursor destined to become either the 28S RNA or 18S RNA components [15].

The '5.8S RNA' fragment appears to be gen-erated simultaneously with the 32S → 28S RNA transition. This has been interpreted by some as indicating the possibility that in the conversion, a loop of 'spacer' RNA located on the precursor chain between the 28S and 5.8S species is degraded.

Whilst in HeLa cells only 56 per cent of the precursor is conserved in the form of mature RNA sequences, the precursor in *Xenopus*, for instance, is smaller (40S, 2.7×10^6 daltons) and almost 80 per cent is conserved. However in the case of *Xenopus* the secondary structure mapping technique has been applied not only to the RNA but to the repeating unit of nucleolar ribosomal DNA. This DNA upon denaturation and appropriate spreading shows a linear arrangement of secondary structure features markedly similar to that in the RNA presursor transcribed from it, as well as regions which are dissimilar and which probably correspond to non-transcribed spacer regions [16]. (See Fig. 3.3). Recent data indicate considerable evolutionary conservation of the sequences that

are transcribed but show heterogeneity in the non-transcribed spacer regions [16]. The role of the spacer regions is not known but it may be that those which are transcribed are concerned with the folding of rRNA precursors or with their interactions with proteins during maturation. In any event it is clear that ribosome manufacture itself is known to take place in the nucleolus. Particulate precursors of cytoplasmic ribosomes ('nascent ribosomal particles') comprising the ribosomal RNA precursors just described, plus newly synthesised ribosomal proteins and the small 5S rRNA, have been isolated from nucleoli. The complex precursor processing steps appear to take place actually within the nascent ribosomal particles themselves (see Fig. 3.4). For a more detailed treatment of this aspect see [17].

For some while it was thought that there was no similar precursor to the ribosomal RNAs in bacteria. However with the use of suitable mutants of *E.coli*, with low levels of RNAse III (see

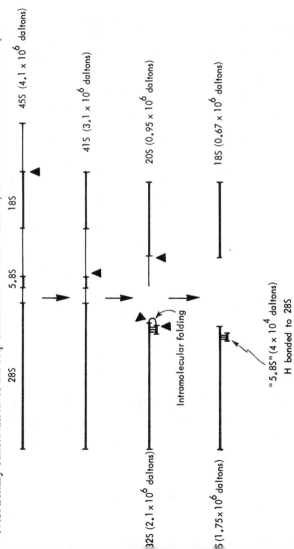

Fig. 3.2 The topographical structure of HeLa cell rRNA precursors and their sites of nucleolar cleavage (▲).

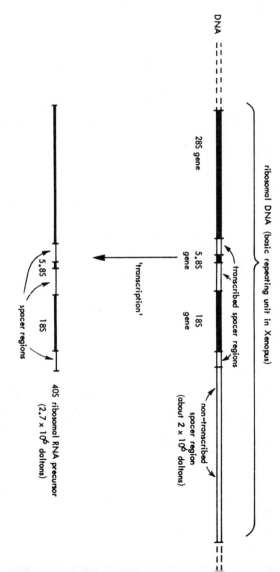

Fig. 3.3 Structure of an rDNA repeating unit in *Xenopus* and its transcription product, the 40S precursor.

Chapter 4) and chloramphenicol treatment, it was possible to observe the transitory appearance of a large 30S transcript containing both 16S and 23S rRNA sequences [18]. Normally, it appears, this RNA is rapidly cleaved during the course of its synthesis to 25S and 17.5S RNA fragments, which in turn are further processed to the mature 23S and 16S ribosomal RNAs of *E.coli* (see Fig. 3.5). Moreover like the situation in higher organisms these RNAs are associated with proteins during processing, to comprise 'nascent ribosomal precursor particles' [19].

3.2.2 *The small 5S ribosomal RNA*

In eukaryotes at least the small 5S ribosomal RNA is not part of the large ribosomal RNA precursor. The genes for 5S rRNA are highly repeated however (5–10 in *E.coli*, 200 in man, and as many as 900–2400 in *Xenopus laevis* [20]), and appear, at least in *Xenopus*, to be tandemly repeated in association with some

'spacer' DNA, but in a number of non-nucleolar loci [20]. In HeLa cells 5S genes are scattered on chromosomes of all sizes [10] and in the anucleolate mutant of *Xenopus* the 5S genes are present in the normal amount whilst the 18 and 28S genes are completely absent. Moreover the transcription of 5S genes can continue when 45S RNA synthesis in mammalian cells is blocked with low levels of actinomycin D [4]. On the other hand in the anucleolate mutant of *Xenopus* mentioned above there is no transcription of the 5S genes that are present, suggesting some 'nucleolar control' over 5S RNA manufacture.

In mammalian cells no precursor to 5S has yet been demonstrated and two types of data suggest that the initial transcript may not be extensively processed, if at all, after transcription. Mature 5S RNA of the ribosomes contains a nucleoside tetraphosphate at its 5′-end [21] characteristic of the 5′-end of a nascent transcript. Kinetic data show label in mature 5S

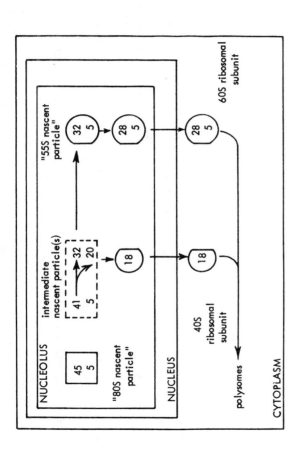

Fig. 3.4 An outline of ribosome formation illustrating the processing of RNA taking place within 'nascent ribosomal particles'. The numbers within the particles refer to the sedimentation coefficient of their constituent RNAs. Note the occurrence of 5S RNA even within '80S nascent particles'.

RNA to appear almost immediately on exposure of cells to isotopically labelled RNA nucleosides [22]. Precursors with only a very few extra nucleotides at the 3'-end cannot yet be ruled out. In fact clear cut evidence is available for such precursors to 5S RNA in *E.coli*. In chloramphenicol treated cells there is an appearance of 5S molecules with one to three extra nucleotides at the 3'-end [23]. The 3'-end of mature 5S rRNA for *E.coli* is normally pU, whereas these precursors have the following 3'-terminal sequences; pUpU, pUpUpU and pUpUpUpA. Processing would simply appear to involve the sequential removal of the extra three nucleotide residues. A longer precursor to 5S RNA has however been reported in *B. subtilis* which has about 30 extra nucleotides at either end.

3.3 Transfer RNA production

As was the case with the ribosomal RNA genes, tRNA genes are reiterated in higher organisms. For instance whereas bacteria possesses 1 or 2 copies of each tRNA gene, yeasts have 5 to 7 copies. *Drosophila* have 13 copies and *Xenopus* and mammalian cells about 200 [24]. These genes like those for 5S rRNA are not located within the nucleolus but are distributed amongst chromosomes of all size ranges [10]. However, recent studies [25] have revealed that the majority of tRNA genes in *Xenopus* are actually neither contiguously linked nor randomly distributed throughout the genome, but are instead clustered with 'spacer' DNA. 'Spacer' DNA segments are on average ten times the size of one tRNA gene (i.e. around 5×10^5 daltons). Specific tDNA 'linkage' groups may

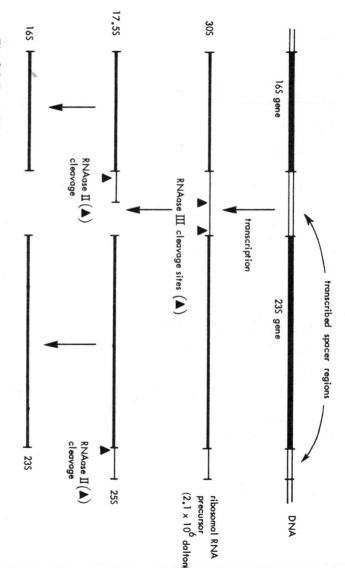

Fig. 3.5 *E.coli* rRNA transcription and processing.

only contain single isocoding genes together with 'spacer' DNA, and may not lie adjacent to other 'tDNA linkage' groups (see Fig. 3.6).

The question that next arises is how are these 'linkage' groups transcribed? Are only the DNA sequences corresponding to the tRNA genes transcribed, or is 'spacer' DNA transcribed as well? There is now a growing body of evidence to suggest that animal and bacterial tRNA's arise not only as a result of 'tailoring' of somewhat longer precursor molecules by scission processes but also by the modification of specific nucleotide sequences.

In eukaryotic cells precursors to tRNA's are rapidly exported to the cytoplasm [4, 24]. These precursors in mammalian cells are longer than tRNA by about 30 nucleotides residues. However, over a period of an hour or so in the cytoplasm, they are cleaved to tRNA dimensions and certain nucleotides are modified by methylation (see Fig. 3.7). In fact there are three different classes of methylation which can be distinguished temporally, (a), early, which appear on or near the junctions of the three

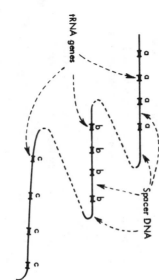

Fig. 3.6 A model for the arrangement of multiple tRNA genes in *Xenopus* [25].

28

major loops, (b), *intermediate*, in the region of the TΨC loop and, (c), *late*, in the region of di-hydrouracil and anticodon loops [26]. Other modifications also occur in the cytoplasm such as the conversion of certain uridine residues to pseudouridine and dihydrouridine residues. The precise structure of these precursors is not yet fully understood mainly because no precursor to a specific mammalian tRNA has yet been isolated from the mixed population. Nevertheless it appears that at least terminal uridine residues must be removed from the 3'-termini during their processing in the cytoplasm.

In *E. coli*, where there is also evidence for precursors of tRNA, the individual genes are not reiterated but are collectively fairly closely clustered [27]. The 'spacer' DNA is apparently quite short (about 30–45 nucleotides in length). A specific study of *E. coli* tyrosine suppressor tRNA and its formation has been made by Smith and his colleagues [28]. A specific precursor to this special tRNA found in the amber suppressor su_3^+ has been isolated and characterised (see Fig. 3.8). Joined to the actual tRNA sequence at its 5'-end are 41 extra nucleotides beginning with pppG and at the 3'-end two extra nucleotides. This precursor is cleaved in *E. coli* to give the mature tRNA sequence. It appears that for processing to be efficient, the tRNA part of the tyrosine tRNA precursor must be folded up in the correct conformation, and that base-substitutions which prevent this, or favour an alternative structure make processing inefficient. Nucleotide modifications apparently take place after the cleavages have occurred.

A novel situation pertains in *E. coli* infected with T4 bacteriophage. Some of the eight tRNA's specifically coded by the T4 genome arise initially in precursors which contain two tRNA sequences (e.g. serine tRNA and proline tRNA). These are subsequently cleaved in the middle to yield two separate tRNA species [29].

3.4 Messenger RNA in prokaryotes

The experiments that led to the establishment of messenger RNA (mRNA) as a real entity were carried out initially on *E. coli* infected with the T-even bacteriophages. Upon infection, bacterial DNA replication stops and the bacterial genome is destroyed. Synthesis of RNA and protein within the infected bacteria continues. In 1957 Volkin and Astrachan [30] determined the base composition of the newly synthesised RNA by transferring the bacteria to a medium containing ^{32}P-orthophosphate, hydrolysing the RNA afterwards, separating the four classes

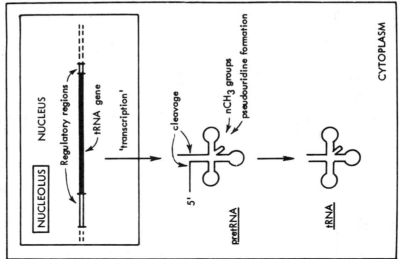

Fig. 3.7 A scheme showing the possible steps involved in the production of mammalian cell tRNA.

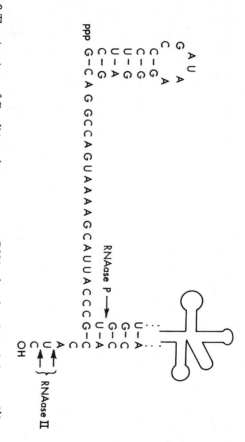

Fig. 3.8 The structure of *E.coli* tyrosine suppressor tRNA, showing sites of cleavage (↑).

of nucleotides and measuring their radioactivity. They found that the newly synthesised RNA had a base composition similar to the bacterio-phage DNA (see Fig. 3.9). This suggested that it had been formed on a DNA 'template'.

Hall and Spiegelman [31] later showed that the RNA synthesised after T2 bacteriophage infection, whilst being relatively metabolically unstable, could form molecular hybrid mole-cules (see Fig. 3.10) with the DNA of the bacteriophage and concluded that the rapidly labelled RNA formed after infection was a T2 specific RNA with a base sequence comple-mentary to the T2-DNA.

The idea that this special RNA formed after infection constituted a messenger between DNA and sites of protein synthesis — as predicted by Jacob and Monod from genetic studies [32] — obtained strong support from several ingenious experiments in which it was shown that this new RNA became attached to pre-existing ribosomes in the bacterial cells [33]. That the ribosomes involved had been made before in-fection was proved by labelling 'heavy' cells with ^{13}C and ^{15}N and infecting *E.coli* cells with T2 bacteriophage in 'light' medium contain-

	A	C	G	U (T)
Normal cells				
DNA	25	25	25	25
rRNA	25	22	31	22
tRNA	21	29	32	19
mRNA	25	24	27	24
Phage-infected cells				
T2 DNA	32	17	18	32
T2 mRNA	30	16	21	22

Fig. 3.9 Molar proportions of bases in the nucleic acids of *E.coli*.

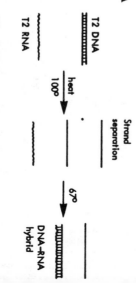

Fig. 3.10 Formation of T2DNA · T2RNA molecular hybrids.

ing ^{12}C and ^{14}N. The T2-specific RNA was found attached to 'heavy' or 'old' ribosomes, as was also the nascent T2-specific protein labelled by pulse exposure to radioactive amino acids. In other words the T2-specific RNA had acted as a messenger from the T2-DNA to the ribosome where it directed the formation of bacterial protein.

From the data in Fig. 3.9 it can be seen that whilst the base composition of the new T2-mRNA is similar to that of T2-DNA, the total content of guanine and cytosine is approximately equal to the corresponding total for the bacteriophage DNA, but the individual bases occur unequally in the RNA (21 per cent for guanine and 16 per cent for cytosine). This observation, suggested Bautz and Hall [34], would be accounted for if the mRNA is always synthesised as the complement of one, rather than both, nucleotide chains of the DNA. Evidence that only one specific DNA chain appears to be employed when a gene functions was also obtained from a study of the effect of 5-fluorouracil on the expression of certain rII mutants of T4 bacteriophage [35].

Additional support for the view that only one strand of the DNA helix is used as template came from studies with other bacteriophages. In bacteriophages α and SP8 the two strands of their DNA differ in density sufficiently for it to be possible to separate the individual 'heavy' and 'light' strands in caesium chloride density gradients. The RNA formed in the bacteriophage infected bacterial cell however forms a molecular hybrid only with one of these strands, the 'heavy' one [36]. Similar studies using separate strands of bacteriophage λ, and with the component strands of bacteriophage φX174 replicative form, led to the same conclusions.

The existence of similar unstable mRNAs in non-infected *E.coli* cells was soon demonstrated [37] both in normal cells, and in 'step-down' cultures in which cells are transferred from a rich medium (permitting rapid growth) to a poor

medium (in which cells contain more ribosomes than necessary). Under these conditions synthesis of ribosomal RNA stops, but since protein manufacture continues at a slow rate some synthesis of mRNA also occurs. On pulse labelling of such cells, a broad peak of radioactivity is found, roughly between 8S and 30S [37]. This material, as anticipated, is metabolically unstable (half-life one to two min) and has a base composition similar to that of *E.coli* DNA (see Fig. 3.9). Also it shows molecular hybrid formation with that DNA but not with bacteriophage DNA or other bacterial DNAs.

This mRNA fraction accounts for one or two per cent of the total RNA. Its molecular weight seems to vary quite considerably and it obviously is quite heterogenous. More advanced techniques combining molecular hybridisation and polyacrylamide gel electrophoresis have however permitted the isolation of specific bacterial messenger such as those specific for the *lac* and *gal* operons. These are polycistronic (see Chapter 2) having molecular weights of 1.7×10^6 [37] and 1.5×10^6 respectively. Study of these has been of obvious use in elucidating the mechanicsm of gene expression at the translational level. Additionally these and other specific polycistronic messengers have yielded valuable information about the metabolic breakdown of messenger after translation. Originally it was thought that bacterial messengers were degraded sequentially in the $5' \to 3'$ direction. An analysis of rates of decay however shows no strong correlation between size and lifetime of these messengers [37, 38]. Furthermore the location of a particular gene proximal or distal to the operator (see Chapter 5) does not appear to have a systematic influence on its rate of decay. It is now felt that the initiation of polycistronic mRNA decay is not progressive from one end of the message but probably starts with specific internal cleavages [37-38]. The cellular site for mRNA degradation may be a ribosomal complex engaged in active protein synthesis.

Having discussed bacterial mRNA formation and its decay it is worth pointing out the fact that so far no post-transcriptional processing has been invoked for normal bacterial mRNAs. It would be premature to rule this out for it may merely have escaped detection. Post-transcriptional processing on the other hand is clearly detectable in the production of the bacteriophage T7-specific 'early' mRNAs that appear in the initial stages of the infectious cycle in *E.coli*. Basically these arise from five cistrons arranged sequentially at the 5'-end of the bacteriophage T7 genome (accounting for 20 per cent of the total genome). These are initially transcribed together as one piece of RNA, however this transcript is quickly cleaved to yield five separate and distinctive monocistronic mRNAs. This is illustrated in Fig. 3.11.

3.5 Messenger RNA formation in eukaryotes

Paradoxically most progress with the isolation of specific mRNAs was made for a while with the mammalian mRNAs (see Chapter 2). The origin of mRNA in eukaryotes still remains somewhat of a puzzle. While it has been known for a long time that a rapidly labelled RNA is formed in the cell nucleus, the possibility of migration of intact polynucleotides from nucleus to cytoplasm has been the subject of much argument.

When mammalian cells for instance are very briefly exposed to ^{32}P-orthophosphate or iso-topically labelled nucleoside precursors the dominant species to be labelled is not the mRNA of the cytoplasmic polysomes, but a polydisperse nuclear RNA fraction which accounts for around 1 per cent of the total cellular RNA. This non-nucleolar nuclear fraction is made up of RNAs whose sedimentation coefficients range widely from 20–100S (i.e. roughly $1–15 \times 10^6$ daltons), hence its name heterogeneous nuclear RNA

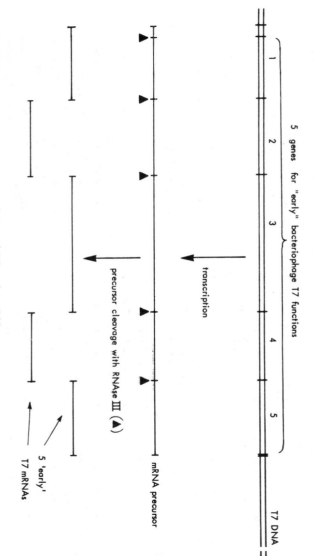

5 genes for "early" bacteriophage T7 functions

1   2   3   4   5

T7 DNA

transcription

mRNA precursor

precursor cleavage with RNAse III (▲)

5' early T7 mRNAs

Fig. 3.11 Post-transcriptional processing of T7 early mRNAs.

32

(hnRNA). Other terms that have been used for this species, which is detectable in the cells of all higher organisms [9,22], are 'DNA-like RNA', or messenger-like RNA'. Basically this is because its base composition is roughly DNA-like, differing markedly from the high proportion of guanine plus cytosine so characteristic of ribosomal RNA. Whereas low levels of actinomycin D will selectively inhibit ribosomal RNA synthesis, specifically 45S RNA synthesis in the nucleolus, hnRNA synthesis continues in the euchromatin regions of the nucleolus. The decay of hnRNA within the nucleus is notably rapid (half life about 20—60 min), but interestingly a proportion (a quarter to a third) have polyA tracts about 200 nucleotides long at their 3'-ends. An mRNA precursor role was postulated on this basis, but experiments designed to follow the metabolic fate of hnRNA molecules led to the conclusion that if there were such a relationship, it was complex. Most of the radioactive label in hnRNA could *not* be traced into the cytoplasm. Nevertheless hnRNA can form molecular hybrids with a large fraction of the genome in higher organisms (up to 11 per cent). The significance of this requires some idea of the complex sequence organisation of the mammalian genome.

In the first place most of the nuclear DNA of eukaryotes is probably non-informational since the DNA content for example of mouse cells is sufficient to code for at least a 1000 times the number of proteins so far found in mice. Secondly an analysis of nucleotide sequences based on the renaturation kinetics of DNA denatured after shearing, indicates some sequences to occur once in the genome and some many times. Many factors of course affect the rate of reannealing or renaturation of the DNA strands (sheared to roughly the same dimensions). These include ionic strength of the medium and the temperature. However, keeping these constant, the rates of DNA renaturation can give a measure of the relative con-

centration in the total nuclear DNA of various nucleotide sequences. A single strand of nuclear DNA which contains sequences that are present in the genome millions of times will clearly find a strand of complementary nucleotide sequence much faster than if it contains a sequence which may only occur once in the genome. The wide variation in rates of reannealing which are found for the sheared DNA of mouse cells after denaturation has been interpreted as reflecting the wide variation in the concentration of various types of nucleotide sequence in the DNA. For instance 10 per cent of the total mouse genome is present as sequences which occur around a million times (highly repetitive, or reiterated DNA). Another 20 per cent is present as sequences which occur 1000 to 100 000 times (intermediate repetitive DNA) and 70 per cent as sequences which occur about once per genome (unique, or non-repetive DNA) [40].

In mouse cells the highly repetitious DNA in fact comprises a family of molecules sufficiently homogeneous and present in sufficiently high relative concentration to give a discrete peak in caesium chloride density gradient analysis. This is termed a 'satellite' peak. Not all eukaryotes have 10 per cent of their genome with this degree of reiteration but all have some highly repetitive sequences. The nucleotide sequences of some nuclear satellites have been determined [41]. So far no proteins have been detected that could have arisen from such sequences. Additionally they do not even appear to be transcribed into RNA, and their function is still not understood. *In situ* molecular hybridisation techniques have however located these repetitive satellite sequences predominantly at the centromeres of chromosomes and also in heterochromatic regions [42].

As regards the intermediate repetitive sequences, these contain the genes for ribosomal RNA, transfer RNA, 5S RNA and histone messenger RNA. Present data however suggest that histone genes are somewhat unusual in that they are reiterated. Most messenger RNAs come from DNA sequences that are represented once per haploid genome, i.e. from unique, or non-repetitive sequences [43].

The arrangement of unique and repetitive sequences in the genome has been studied in some eukaryotes and the results indicate that generally the intermediate repetitive and unique DNA sequences are not clustered like the highly repetitive sequences. Instead they are interspersed with one another. In half the genome of *Xenopus* for instance intermediate repetitive sequences are present as average lengths of 200-400 nucleotides, alternating with non-repetitive sequences of an average length of around 650-900 nucleotides. Another quarter of the genome may be organised in a longer period interspersion with intermediate repetitive sequences, at present of undefined length, separated by non-repetitive sequences of from 4000-8000 nucleotides. Not more than 6-8 per cent seems to comprise clustered intermediate repetitive sequences [44]. A similar situation pertains in the sea urchin genome [45].

Whilst most mRNA [43] would appear to have arisen as the result of transcription of unique DNA sequences, the same is not true for its putative nuclear precursor, hnRNA. Covalently attached to one another in hnRNA molecules are sequences which are transcripts of both unique and intermediate repetitive sequences. The interspersion and length of these sequences in the hnRNA reflect to some extent the length and interspersion of the corresponding sequences in the nuclear DNA [47]. Additionally the structural features of the HeLa cell hnRNA molecules that terminate in a poly A tract, have been probed by fragmenting the molecules by mild alkaline treatment. HnRNA fragments of 3000 nucleotides (approximate mRNA dimensions) containing the 3'-terminal poly A tract hybridise to HeLa DNA at the same rate as mRNA (i.e. they originate from unique sequences in the genome). The remaining 5'-portion of the hnRNA (see Fig. 3.12) molecules however, contain [41] (a), intermediate repetitive and unique sequences interspersed with one another, (b), two or three short tracts of about 30 uridylate nucleotides each (oligoU tracts) [48], (c), double-stranded regions, and (d), possibly oligoA tracts of 30 units (see Fig. 3.12). Despite this structural information, a precursor-product relationship with mRNA has been difficult to establish unambiguously. Is the 3'-end of hnRNA with the unique sequence conserved during processing and exported to the cytoplasm? Is the polyA conserved?

Regarding the first of these questions there are some data on sequence similarly between hnRNA and mRNA. For example, sequences complementary to certain viral DNAs (SV40,

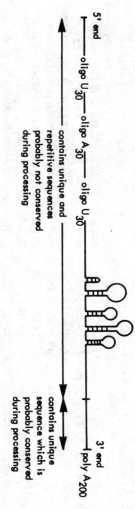

5' end

oligo U $_{30}$ — oligo A $_{30}$ — oligo U $_{30}$

contains unique and repetitive sequences probably not conserved during processing

3' end poly A $_{200}$

contains unique sequence which is probably conserved during processing

Fig. 3.12 Some structural features of hnRNA from HeLa cells. The arrangement of the oligoA, oligoU and double-stranded regions towards the 5'-end is highly schematic, their precise locations have yet to be determined.

adenovirus, and herpes) have been found to appear in hnRNA and then in mRNA [49]. Additionally, new techniques whereby highly labelled DNA is made complementary to purified mRNAs with oncornavirus 'reverse transcriptase' (dealt with in T.H. Pennington and D.A. Ritchie, *Molecular Virology*, in this series) and used to probe nuclear RNAs for messenger-like sequences, have detected globin mRNA sequences in duck reticulocyte hnRNA [50] and ovalbumin mRNA sequences in chick oviduct hnRNA [51].

Most present day data support the view that the formation of polyA takes place by the sequential addition of adenylate moieties to completed RNA molecules by a process independent of a DNA template. The process is insensitive to actinomycin D, and whilst there are poly dT tracts in mammalian DNA they are insufficiently long (20–40 nucleotides) to code for the polyA tracts (200 long) in hnRNA and mRNA [52]. Moreover a nuclear enzyme capable of carrying out this sequential addition has been detected in nuclear ribonucleoprotein particles [53,54] (see Chapter 4). In HeLa cells exposed to labelled adenosine, the proportion of labelled adenosine in polyA tracts is greater in those from hnRNA than in those from mRNA after short labelling periods (15–25 min). The situation is however reversed after longer labelling periods (150 min), consistent with a precursor-product relationship. Is this however connected with mRNA appearance in the cytoplasm? Here, studies with the drug 3'-deoxyadenosine (cordycepin) have been useful. Whilst the appearance of mRNA is inhibited by this drug, hnRNA synthesis is relatively unimpaired [55]. The explanation for this that is in current vogue, is that the drug specifically inhibits the addition fo the polyA tracts to hnRNA molecules. This event may be in some complex way involved in selecting hnRNA molecules for processing to yield cytoplasmic mRNA terminated with polyA

tracts. If HeLa cells are exposed to labelled nucleoside for a short period (7.5 min) and then transcription stopped with actinomycin D, mRNA normally appears in polysomes some 20 min later. 3'-deoxyadenosine however blocks the entry of 70 per cent of the mRNA into the polysomes and those that do arrive have only very short polyA tracts [56]. Whether or not polyA addition is required for mRNA export has to be considered in the light of data indicating histone mRNA to enter polysomes with only a delay of 5 min. Taken together with its known lack of a polyA tail, this suggests that dispensing with polyA addition may permit more rapid processing [56]. Indeed a sizeable proportion of mRNAs in both HeLa cells [57], and sea urchins [65] now appears not to have polyA tracts, so leaving the function of polyA an open question. Furthermore, recent data that indicate a metabolic turnover of the polyA tracts themselves in the nucleus as well as in the cytoplasm, serve to confirm the complexity of the sitation and make kinetic analysis difficult [58].

In summary therefore it does seem likely that the hnRNA and mRNA are related and that hnRNA does contain mRNA precursors. Clearly a number of questions remain to be answered. What is the exact structure of hnRNA, the nature and function of its non-coding regions, the nature of enzymes invloved in its cleavage, the proportion that contain mRNA sequences and the number of these per molecule? Finally it should be pointed out that all the events discussed above occur not on nuclear RNA molecules but more likely on these RNAs as part of complexes with protein (ribonucleoprotein complexes). A variety of complexes of protein with hnRNA and with mRNA (mRNP) have been detected by various workers in the nucleus and cytoplasm of a number of animal cells. However there is, as yet, only imprecise data on the relationship between these various complexes and many problems arise from artefacts. It is clear that a full understanding of mRNA production and transport from the nucleus will require an appreciation of the role of these protein complexes. For a full appreciation of the present situation the reader is referred elsewhere [59].

A final comment on eukaryotic mRNA biosynthesis concerns a very recent finding that the 5′-terminal sequence of a wide variety of messengers is quite unusual. It consists of 7-methylguanosine linked through its 5′-hydroxyl, via a tri (or pyro)-phosphate group, to the 5′-hydroxyl of 2′-methylated nucleoside [62,63] (see Fig. 3.13). Removal of the 7-methylguanosine moiety from this sequence is sufficient to prevent the translation of the mRNA [63]. The origin of this very novel sequence, or 'cap' as it is sometimes called, is not yet clear. A suggestion is that 'capping' of the 5′-terminus is a post-transcriptional process. The guanosine may be added enzymically to the 5′-end on the new mRNA molecules from GTP as substrate and its subsequent methylation may involve a methylase and S-adenosyl-L-methionine (see Chapter 4). Besides this unusual methylated sequence there are other methylated nucleotides in mammalian mRNA. However they only occur at a low level and their precise location is not yet known [64].

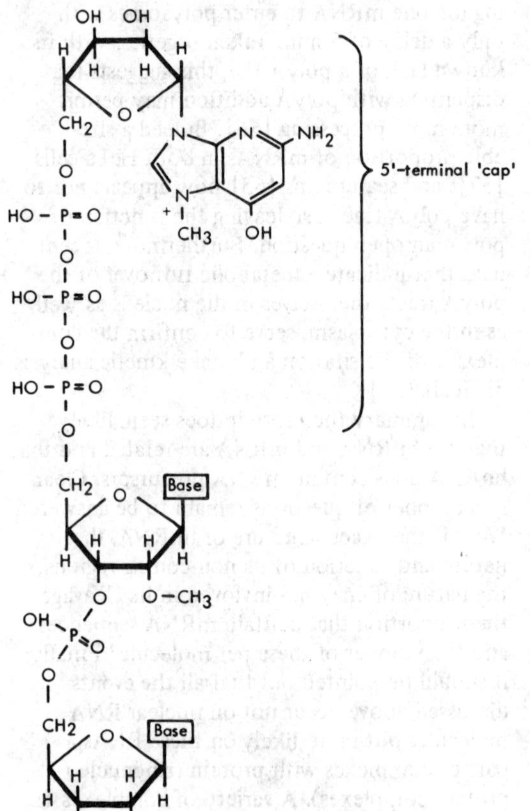

3.6 Patterns of mitochondrial transcription

The individual strands of the cyclic DNA duplexes that comprise the genome of mitochondria can be conveniently separated from one another by density gradient techniques. These single strands are termed the 'heavy' (H) and 'light' (L) strands [60]. Molecular hybridisation [61] in the case of HeLa cells showed that the mitochondrial ribosomal RNAs (12S and 16S) are transcribed from the H strand. Also on the H strand are nine sites for tRNA transcription but a further three sites seem to be on the other strand, the L strand (see Fig. 3.14). These 'genes' account for around 25 percent of the potential information. However this does not take account of the intriguing possibility of meaningful information being coded on *both strands at the same site*. Indeed it has been shown that both strands are in fact almost completely transcribed. In cells exposed to labelled uridine for long periods (46 hr), the mitochondrial RNA so labelled, hydridised almost exclusively to the H strand. The small amount

5′-terminal 'cap'

Fig. 3.13 The 5′ terminal structure of eukaryotic mRNA.

36

which hybridised to the L strand appeared to include that hybridising to the three tRNA genes. The situation after very short 'pulse labels' was quite different. There seemed to be equal hybridisation of the mitochondrial RNA to both strands [61].

It may be that HeLa cell mitochondria solve the problem of strand selection in transcription by transcribing both strands and rapidly degrading 98 per cent of the L-strand transcripts. This of course poses questions as to how the tRNA or rRNA (or mRNA) sections of these transcripts are correctly cut out. Whilst the

bromide, a drug which specifically intercalates with the supercoiled cyclic DNA molecules of the mitochondria. Analysis of this polyA-terminated mitochondrial polysomal RNA by polyacrylamide electrophoresis showed it to occur as two distinct components, a 9S component which is coded for by the H-strand, and a 7S component coded for mainly by the L-strand [61]. This latter observation adds at least one more RNA species to the three transfer RNA species already known to be coded for by this L-strand.

main features of gene distribution over the two strands seems also to apply to rat and *Xenopus* mitochondrial DNA, it is not yet clear whether the same pattern of transcription occurs.

Regarding mRNAs in mitochondria, polyA added post-transcriptionally has been detected in the RNA from mitochondrial polysomes. Its length however is smaller than that on cytoplasmic mRNA, being about 60—80 nucleotides in length [61]. That the polysomal RNA is a mitochondrial DNA transcript is suggested by the fact that it hybridises to mitochondrial DNA, and its production is inhibited by ethidium

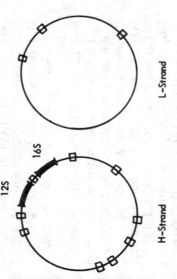

Fig. 3.14 Circular map of the positions of sequences coding for rRNAs (□) and 4S RNAs (□) on H and L strands of HeLa mitochondrial DNA.

References

[1] Perry, R.P. (1967), *Prog. Nucleic Acid Res. Mol. Biol.* **6**, 219.

[2] Weiss, S.B. (1960), *Proc. Nat. Acad. Sci. U.S.* **46**, 1020.

[3] Burdon, R.H. and Smellie, R.M.S. (1962) *Biochim. Biophys. Acta*, **61**, 633.

[4] Burdon, R.H. (1971), *Prog. Nucleic Acid Res. Mol. Biol.* **11**, 33.

[5] Bishop, J.O. (1972), *Fifth Karolinska Symposium 'Gene transcription in reproductive tissue'*, Ed. Diczfalusy, E., Karolinska Institute, Stockholm, p. 247.

[6] Kennell, D.E. (1971), *Prog. Nucleic Acid Res. Mol. Biol*, **11**, 259.

[7] Yanofsky, S.A. and Spiegelman, S. (1962), *Proc. Nat. Acad. Sci, U.S.*, **48**, 1466.

[8] Maden, B.E.H. (1971), *Prog. Biophys. Mol. Biol*, **22**, 127.

[9] Darnell, J.E. (1968), *Bact. Rev.*, **32**, 262.

[10] Evans, H.J. (1973), *Brit. Med. Bull.*, **29**, 196.

[11] Birnstiel, M.L., Wallace, H., Sirlin, J.L. and Fischberg, M. (1966), *Nat. Cancer. Inst. Monograph*, **23**, 431.

[12] Birnstiel, M.L., Spiers, J., Purdon, I., Jones, K. and Loening, U.E. (1968), *Nature*, **219**, 459.

[13] Penman, S., Smith, I. and Holtzman, E. (1966), *Science*, **154**, 789.

[14] Wellauer, P.K. and Dawid, I.B. (1973),

Proc. Nat. Acad. Sci., U.S., **70**, 2827.

[15] Maden, B.E.H., Salim, M. and Shepherd, J. (1973), *Biochem Soc Symp*, **37**, 23.

[16] Wellauer, P.K. and Dawid, I.B. (1974), *J. Mol. Biol.*, **89**, 379.

[17] Perry, R.P. (1973), *Biochem. Soc. Symp.*, **37**, 105.

[18] Nikolaev, N., Schlessinger, D. and Wellauer, P.K. (1974), *J. Mol. Biol.*, **86**, 741.

[19] Pace, N.R. (1973), *Bact. Rev.*, **37**, 562.

[20] Ford, P.J. (1973), *Biochem. Soc. Symp.*, **37**, 69.

[21] Hatlen, L.E., Amaldi, F. and Attardi, G. (1969), *Biochemistry*, **8**, 4989.

[22] Weinberg, R.A. (1973), *Ann. Rev. Biochem.*, **42**, 329.

[23] Forget, B.G. and Jordan, B. (1970), *Science*, **167**, 382.

[24] Burdon, R.H. (1975), *Brookhaven Symposium in Biol.*, **26**, (in press).

[25] Clarkson, S.G., Birnstiel, M.L. and Serra, V. (1973), *J. Mol. Biol.*, **79**, 391.

[26] Munns, T.W. and Katzman, P.A. (1973), *Biochem. Biophys. Res. Commun.*, **53**, 119.

[27] Fournier, M.J., Miller, W.L. and Doctor, B.P. (1974), *Biochem. Biophys. Res. Commun.*, **60**, 1148.

[28] Smith, J.D. (1973), *Brit. Med. Bull.*, **29**, 220.

[29] Guthrie, C., Seidman, J.G., Altman, S., Barrell, B.G., Smith, J.D. and McClaine, W.H. (1973), *Nature*, **246**, 6.

[30] Volkin, E. and Astrachan, L. (1957), in 'The chemical basis of heredity.' Eds. McElroy, W.D. and Glass, B., Johns Hopkins Press, Baltimore p. 686.

[31] Hall, B.D. and Spiegelman, S. (1961), *Proc. Nat. Acad. Sci, U.S.*, **47**, 137.

[32] Jacob, F. and Monod, J. (1961), *J. Mol. Biol.*, **3**, 318.

[33] Brenner, S., Jacob, F. and Meselson, M. (1961), *Nature*, **190**, 576.

[34] Bautz, E.K.F. and Hall, B.D. (1962), *Proc. Nat. Acad. Sci.*, **48**, 400.

[35] Champe, S.P. and Benzer, S. (1962), *Proc. Nat. Acad. Sci., U.S.*, **48**, 332.

[36] Marmur, J., Greenspan, C.M., Policek, E., Kahan, F.M., Levine, J. and Mandel, M. (1963), *Cold Spring Harbor Symp. Quant. Biol.*, **28**, 191.

[37] Blundell, M. and Kennell, D. (1974), *J. Mol. Biol.*, **83**, 143.

[38] Achord, D. and Kennell, D. (1974), *J. Mol. Biol.*, **90**, 581.

[39] Dunn, J.J. and Studier, W.F. (1973), *Proc. Nat. Acad. Sci., U.S.*, **70**, 1559.

[40] Britten, R.J. and Kohne, D.E. (1968), *Science*, **161**, 529.

[41] Southern, E.M. (1970), *Nature*, **227**, 794.

[42] Jones, K.W. (1970), *Nature*, **225**, 912.

[43] Bishop, J.O., Morton, J.G., Rosbash, M. and Richardson, M. (1974), *Nature*, **250**, 199.

[44] Davidson, E.H., Graham, D.E., Neufeld, B.R. Chamberlin, M., Amenson, C.S., Hough, B.R. and Britten, R.J. (1973), *Cold Spring Harbor Symp. Quant. Biol.*, **38**, 295.

[45] Graham, D.E., Neufeld, B.R., Davidson, E.H. and Britten, R. (1974), *Cell*, **1**, 127.

[46] Klein, W.H., Murphy, W., Attardi, G., Britten, R.J. and Davidson, E.H. (1974), *Proc. Nat. Acad. Sci., U.S.*, **71**, 1785.

[47] Molloy, G.R., Jelinek, W., Salditt, M., Darnell, J.E. (1974), *Cell*, **1**, 43.

[48] Burdon, R.H. and Shenkin, A. (1972), *FEBS Letts*, **24**, 11.

[49] Darnell, J.E., Jelinek, W.R. and Molloy, G.R. (1973), *Science*, **181**, 1215.

[50] Macnaughton, M., Freeman, K.B. and Bishop, J.O. (1974), *Cell*, **1**, 117.

[51] McKnight, G.S., Schimke, R.T. (1974), *Proc. Nat. Acad. Sci., U.S.*, **71**, 4327.

[52] Shenkin, A. and Burdon, R.H. (1974), *J. Mol. Biol.*, **85**, 19.

[53] Edmonds, M. and Abrams, R. (1960), *Biol. Chem.*, **235**, 1142.

[54] Burdon, R.H. (1963), *Biochem. Biophys. Res. Comm.*, **11**, 472.

[55] Penman, S., Rosbash, M. and Penman, M. (1970), *Proc. Nat. Acad. Sci., U.S.,* **67,** 1878.

[56] Adesnick, M., Salditt, M., Thomas, W. and Darnell, J.E. (1972), *J. Mol. Biol.,* **71,** 21.

[57] Milcareck, C., Price, R. and Penman, S. (1974), *Cell,* **3,** 1.

[58] Perry, R.P., Kelley, D.E. and La Torre, J. (1974), *J. Mol. Biol.,* **82,** 315.

[59] Williamson, R. (1973), *FEBS Letts.,* **37,** 1.

[60] Borst, P. (1972), *Ann. Rev. Biochem.,* **41,** 333.

[61] Attardi, G., Constantino, P. and Ojala, D. (1974), in *The Biogenesis of Mitochondria* Eds. Kroon, A.M. and Saccone, C. Academic Press, London, p. 9.

[62] Adams, J.M. and Cory, S. (1975), *Nature* **255,** 28.

[63] Muthukrishnan, S., Both, G.W., Furuichi, Y. and Shatkin, A. (1975), *Nature,* **255,** 33.

[64] Perry, R.P. and Kelley, D.E. (1974), *Cell,* **1,** 37.

[65] Nemer, M. (1974), in *Concepts of Development* Eds. Lash, J. and Whittaker, J.R., Sinauer Associates Inc. p. 119.

4 The enzymic machinery

4.1 Transcription apparatus

In the process of DNA transcription the positioning of nucleotide units in the RNA molecules being made is under the control of the DNA, which acts as template. The means by which this template dictates such a sequence involves both base pairing interactions and specific interactions between proteins and nucleic acids. Additionally, each RNA chain is initiated at a specific site on the DNA template and subject to termination at another unique type of site on the template. In other words, there are defined units of transcription. It is a selective process. Specific signals in the DNA template are recognized by the transcription apparatus. Initiation is governed by *promoter* regions in the DNA, and a region governing termination is designated a *terminator*.

Transcription is mediated by DNA-dependent RNA polymerases, which have now been isolated from a wide variety of sources, eukaryotic and prokaryotic. The properties of the enzyme from *E.coli* have however been the most widely explored. This purified enzyme can carry out the selective transcription of certain DNAs *in vitro*.

4.1.1 The mode of action and structure of
of E.coli RNA polymerase

The first indication of the existence of a DNA-dependent RNA synthesising enzyme in bacterial cells using the four ribonucleoside 5'-triphosphates as substrates came to light in the early 1960s. Basically, the reaction catalysed in the test-tube by the enzyme can be represented as follows (Fig. 4.1). The actual nucleo-

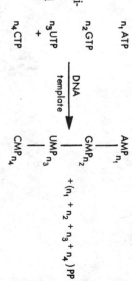

$$
\begin{array}{c}
n_1\,\text{ATP} \\
n_2\,\text{GTP} \\
n_3\,\text{UTP} \\
+ \\
n_4\,\text{CTP}
\end{array}
\xrightarrow{\text{DNA template}}
\begin{array}{c}
\left.\begin{array}{c} \text{AMP}\,n_1 \\ \text{GMP}\,n_2 \\ \text{UMP}\,n_3 \\ \text{CMP}\,n_4 \end{array}\right\} + (n_1 + n_2 + n_3 + n_4)\,\text{PP}
\end{array}
$$

Fig. 4.1 The basic RNA polymerase reaction.

tide composition of the RNA produced was soon shown to be determined by the nature of the DNA, added as template. Moreover the technique of 'nearest neighbour analysis' provided compelling evidence that the added DNA template also determined the sequence of the nucleotides in the RNA made [1]. This aspect was amply confirmed by the then newly developed technique of molecular hybridisation. For example, radioactive RNA prepared *in vitro* using T2 bacteriophage DNA as template was capable of forming a stable RNA-DNA hybrid with denatured T2 bacteriophage DNA. In other words, the RNA made in the test-tube had a nucleotide sequence precisely complementary to a strand of the DNA which was added to the reaction mixture as template [2].

The DNA-dependent RNA polymerase has now been extensively purified from *E.coli*. The holoenzyme is a complex zinc-containing protein of molecular weight 480 000–500 000 which can be dissociated, for instance in 6M urea, into a number of polypeptide chain subunits as follows:

Two α chains each of mol.wt 39 000
One β chain of mol.wt 155 000
One β' chain of mol.wt 165 000
One molecule of σ factor, mol.wt 95 000.

The holoenzyme may therefore be represented as $\alpha_2\beta\beta'\sigma$. The holoenzyme without σ factor is termed 'core' enzyme, or $\alpha_2\beta\beta'$. Chromatography on phosphocellulose was used by Burgess and his colleagues [3] to separate the holoenzyme into the 'core' enzyme and the σ factor. The 'core' enzyme will catalyse the synthesis of RNA chains from random sites on a DNA template in vitro quite well when 'foreign' DNA (for instance that from calf thymus) is added, but ineffectively when the DNA added is that from E.coli or T4 bacteriophage. Addition of σ factor however restores synthetic activity of the 'core' enzyme by activating selective initiation of RNA synthesis [3]. It complexes with the 'core' enzyme to yield holoenzyme which interacts specifically with promoter regions on the DNA template. Using T4 bacteriophage DNA as in vitro template, the σ factor addition specifically stimulates the 'core' enzyme into transcribing those bacteriophage genes which correspond precisely to those normally expressed in vivo during the early stages of T4 bacteriophage infection of E.coli.

The sequence of molecular events for accurate initiation of RNA synthesis can be envisaged as follows [4]. After a series of unproductive random interactions with the DNA template, the holoenzyme 'recognises' a specific structure, or sequence, within a promoter region. This enables the holoenzyme to bind to the DNA in this region at least an order of magnitude more tightly than in the non-specific interactions mentioned above. The precise structural features of the DNA actually required here are not yet known however the sequence

[5] of the 'promoter' region for the E.coli lac operon is shown in Fig. 4.2.

The next step in the initiation process is a change in the conformational state of the DNA in this promoter. Once this transition has occurred the holoenzyme is poised to commence the manufacture of an RNA chain (see Fig. 4.3), but actually using only one of the strands as template. The nature of the conformation change is thought [6] to involve a local unwinding or strand separation of the DNA double helix over 4 to 8 base pairs.

Although it is the β'-subunit that is implicated in the binding of the polymerase to the DNA, and perhaps has the required specificity, σ appears to bind to the β-subunit [7]. It may be best to consider the σ factor as an allosteric effector of a multisubunit enzyme. The β-subunit itself is probably most concerned with at least the initial catalytic reactions i.e formation of the first internucleotide linkage of an RNA chain. These initial reactions (as distinct from later chain elongation reactions) can be blocked by the antibiotic rifampycin, and genetic data indicate that rifampycin resistant mutants carry the resistant phenotype in this β-subunit [8]. The means whereby rifampycin blocks initiation is not yet clear, but it does appear to interfere with the binding of the polymerase substrates GTP and ATP to the β-subunit [9]. Such nucleoside triphosphates are commonly found at the ends of new RNA chains made by the holoenzyme e.g. pppGpYp ... and pppApYp ... etc. [10]. Once the initial internucleotide link between either GTP or ATP and the next nucleotide specified by the template has been made, the σ factor is released, thus possibly reducing the affinity of the polymerase for the promoter site and allowing the remaining 'core' to move along the DNA template spinning off a strand of RNA as it proceeds.

Basically the new chain grows by the subsequent addition of ribonucleotides to the free 3'-hydroxyl group [10]. For instance the RNA

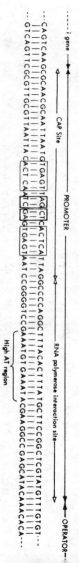

```
···CAGTCAAGCGCAAACGCAATTAATGTGAGTTAGCTCACTCATTAGGCACCCCAGGCTTTACACTTTATGCTTCCGGCTCGTATGTTGTGT···
···GTCAGTTCGCGTTTGCGTTAATTACACTCAATCGAGTGAGTAATCCGGGGTCCGAAATGTGAAATACGAAGGCCGAGCATACAAACACA···
```

Fig. 4.2 The *lac* promoter sequence showing RNA polymerase interaction site (for significance of CAP binding site see Chapter 5). Boxes indicate regions of symmetry.

(a)
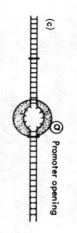

RNA polymerase

Promoter

(b)

Promoter recognition

(c)

Promoter opening

(d)
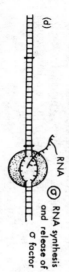

RNA

RNA synthesis and release of σ factor

Fig. 4.3 A diagrammatic representation of the stages in initiation of RNA synthesis.

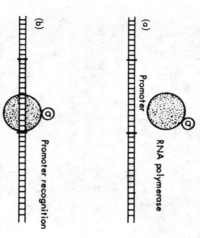

being made in Fig. 4.4 has a triphosphate group at position 5′ on the first nucleotide and a free hydroxyl group at position 3′ at the other,

or growing end. Alkaline hydrolysis of this particular RNA [10] will yield a molecule of the nucleoside cytosine (C) from the 3′-hydroxyl end of the molecule, uridine and adenosine monophosphates (Up and Ap) and a molecule of guanosine tetraphosphate (pppGp) from the 5′-end. Synthesis proceeds from the 5′-end to the 3′-end of the RNA molecule. This can also be shown to be the case *in vitro*. Still unclear are the precise characteristics of DNA structure at the *promoter*, where the DNA is actually used as template rather than just for the tight binding of the RNA polymerase. In any event, the nucleotide residue at the 5′-end of the nascent chain always contains a purine base, thus any DNA sequence involved in the actual commencement of template directed RNA synthesis must contain a thymine or cytosine nucleotide.

Whilst the antibiotic rifampycin only blocks initiation of new chain synthesis, the drug actinomycin D selectively prevents elongation without affecting polymerase-DNA binding, or initiation, by complexing to deoxyguanosine residues on the DNA template and so preventing the movement of the 'core' along the template [12].

The selective termination of RNA chains *in vitro* is less well understood. It appears in the first place that it can be terminated directly by virtue of specific sequences on the DNA template acting directly on the 'core' enzyme. In principle the process of termination involves,

(a) the cessation of RNA chain elongation,
(b) the release of the newly formed RNA, and
(c) the release of the RNA polymerase from the

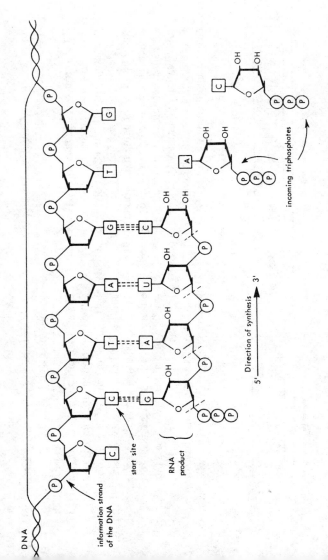

Fig. 4.4 A diagrammatic representation of the biosynthesis of RNA on one strand of DNA acting as template. (The dotted lines indicate sites of hydrolysis with alkali).

DNA. All three of these reactions can occur at 'direct terminator' sites *in vitro*. Little is known regarding the structure of such sites but in certain cases (e.g. the 'early' *in vivo* transcript of T7 bacteriophage (see Fig. 3.11)) specific 3'-terminal nucleotide sequences have been identified. A second type of termination can be induced *in vitro* by a factor, called rho (ρ) [13]. This factor from *E.coli* is an oligomeric protein, (monomer molecular weight 50 000) which depresses the amount of RNA formed in the *in vitro* reaction by causing RNA chain termination. The resulting RNA chains can be released from the DNA but the RNA polymerase remains bound to the DNA. Thus ρ-factor by itself does not allow continued recycling of RNA polymerase through the transcription sequence and a reconstruction of the physio-

logical sequence may require some additional component for the termination-release step. Highly purified ρ-factor preparations contain components with a hexagonal subunit structure, some of which can bind to certain DNAs [14]. This binding ability may be related to its ability to terminate transcription. In any event it appears at the moment that ρ-factor at low concentrations may cause termination at certain *terminators* used *in vivo*, higher concentrations can cause termination at a variety of sites on DNA some of which are not biologically active *terminators*.

(As already mentioned, relatively little is known about the structure of *terminators*, however by *in vitro* chemical and enzymic methods [41] some information has been gained about the nucleotide sequences of a possible *termina-

43

tor region of the *E.coli* tyrosine suppressor tRNA gene (see Fig. 4.5.)

4.1.2 The eukaryotic nuclear DNA-dependent RNA polymerases

In mammalian cells the RNA polymerase activity detected in nuclei was initially difficult to study by virtue of being tightly bound to the nuclear chromatin complex [15,16]. Somewhat specialised techniques were required for its solubilisation in reasonable yields before purification, similar to that achieved for the *E.coli* enzyme, could even commence.

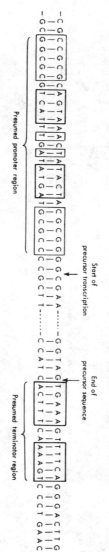

Presumed promoter region

Start of precursor transcription

End of precursor sequence

Presumed terminator region

Fig. 4.5 Nucleotide sequences adjacent to *E.coli* tyrosine suppressor tRNA precursor gene [41]. Boxes indicate regions of symmetry.

Depending on the origin of the cells used, different approaches to solubilisation were employed. Some procedures were mild and involved merely incubating, or homogenising nuclei, in slightly alkaline buffer. A moderately drastic treatment involved brief sonication in low ionic strength medium. In a more drastic approach, sonication was carried out in a medium of high ionic strength [17].

After solubilisation from chromatin, chromatography of the enzyme activity from a variety of animal tissues on DEAE-cellulose, or DEAE-Sephadex, revealed the presence of multiple forms of RNA polymerase. This multiplicity of RNA polymerase has also been observed in lower eukaryotes such as an aquatic fungus, yeast and maize. The enzyme activities are usually eluted from the columns using a linear gradient of ammonium sulphate, or KCl. Depending on the tissue used, two to

three discrete peaks of activity can be resolved by this approach [17].

The classes of RNA polymerase separated chromatographically between 0.10 and 0.37M salt concentrations are referred to usually as I, II and III, in order of their elution. To add another level of complexity, classes I, II and III, which are the major species detected in most eukaryotic cells, have each been further resolved into at least two classes (e.g. III$_A$ and III$_B$ etc.)

Class I RNA polymerases have been established to be of nucleolar origin, whereas classes II and III are of nucleoplasmic origin [17,18]. Additionally these enzymes operate optimally under somewhat different conditions. The nucleolar enzymes work best at low ionic strength and utilize Mn^{2+} and Mg^{2+} equally well. On the other hand, higher Mn^{2+} concentrations and ionic strengths are required for maximum activity of the nucleoplasmic enzymes. Moreover the activity of nucleoplasmic class II enzymes are inhibited selectively by α-amanitin (a toxin from the poisonous mushroom, *Amanita phalloides*) at concentration as low as 3 × 10^{-8} M, whereas the nucleolar activity is not affected even at much higher doses [17]. The minor nucleoplasmic class III enzymes also are not affected by α-amanitin [19].

α-amanitin is specific for the eukaryotic polymerases of class II, as rifampycin is specific for the prokaryotic enzyme. The mushroom toxin appears to bind to the RNA polymerase rather than to the DNA template like actino-mycin D. Whereas rifampycin inhibits initiation by the bacterial polymerase, α-amanitin blocks RNA synthesis after initiation, presumably at the level of chain elongation. In fact its action resembles more the action of another bacterial RNA polymerase inhibitor, streptolydigin, which has been recently shown to block elong-ation in bacteria.

Since factors such as σ and ρ appear to play an important role in the control of transcript-ion a search for such factors was made in mam-malian systems. A protein factor (mol.wt. 70 000), which can specifically bind to and stimulate, the activity of nucleoplasmic class II enzymes from rat liver and calf thymus, has been tentatively identified in the cytoplasm of rat liver [20]. However the question whether or not the factor can promote initiation of RNA synthesis at specific sites on the DNA template cannot yet be answered. With regard to termination in eukaryotes no information is presently available.

Having mentioned a possible factor, another question is how do these enzymes relate struct-urally to the prokaryotic enzyme described previously (4.1.1)? The situation is complex but structural analyses [21,22] have shown that the molecular weights of the large subunits detected in the Class III enzymes (138 000 and 155 000) differ from those of the class II en-zymes (140 000 and either 170 000, or 205 000, or 240 000) and from those of the class I en-zymes (117 000 and 195 000). Some low molecular weight subunits are also unique to each enzyme class. These data clearly distinguish class I, II and III enzymes on a structural basis. In addition polypeptides of molecular weight 29 000 and 19 000 were found in all classes, a polypeptide of molecular weight 52 000 was found only in class I and III enzymes, and a polypeptide of molecular weight 41 000 was found only in class II and III enzymes. Thus it appears from these findings, and the different

sensitivities to rifampycin and α-amanitin, that there are no real structural similarities between the eukaryotic polymerases and the 'core' poly-merase of prokaryotes. Additionally, it appears that the three classes of polymerase are assembled primarily from distinct gene products,, and that they are not interconvertible by simple structural alterations.

Despite these advances, the role of these various polymerases in cellular RNA synthesis remains to be clarified. At least in E.coli, two types of data support the view that the E.coli enzyme studied in vitro, is also responsible for all the E.coli RNA synthesis in vivo: (a) that certain temperature sensitive mutants in RNA synthesis can be shown to have extremely heat labile RNA polymerase, and (b), that rifampycin, which blocks the initiation of all cellular RNA synthesis, can be shown to bind to the β-subunit, and that resistant mutants carry the resistant phenotype in the same β-unit.

In eukaryotes there is no similar direct genetic evidence. It is quite possible that the nucleolar enzymes (I) are involved in the synthesis of the ribosomal precursor RNA (e.g. the 45S of mammalian cells) whereas the nucleoplasmic enzymes (II and III) make hnRNA (or mRNA), tRNA and 5S RNA. Indeed product analysis indicates enzyme III to be specifically res-ponsible for the biosynthesis of 5S RNA and the precursor to tRNA in mammalian cells [19], thus leaving the possibility that enzyme II is involved in hnRNA and/or mRNA product-ion.

4.1.3 The transcription apparatus of mitochon-dria

Compared with the nuclear DNA-dependent RNA polymerases relatively little is known about the properties and functions of the mi-tochondrial polymerases. Soluble preparations have now been obtained from a number of sources and the enzymes from *Neurospora crassa* and yeast have been extensively purified

[23,24]. That from rat liver appears closely associated with the mitochondrial membrane and detergents are required for its extraction [25]. Like the prokaryotic RNA polymerases the enzymes from rat liver, heart or *N. crassa* are sensitive to rifampycin. α-amanitin on the other hand has no effect on the mitochondrial enzymes. Despite this similarity with the bacterial polymerase, the mitochondrial enzymes seem to comprise single subunits of relatively low molecular weight (64 000–68 000) [17].

4.2 Enzymic machinery of post-transcriptional processing

Basically there are three classes of enzymes involved, (a) those that add nucleotides terminally to preexisting RNA molecules (b) those that modify the nucleotide sequence of newly synthesised RNAs (e.g. those that methylate specific nucleotide residues) and (c) those that cleave nascent RNA molecules to smaller products.

4.2.1 The terminal adding enzymes

The covalent attachment of polyA to the 3'-ends of hnRNA and/or mRNA (see Chapter 3) may well be catalysed by a terminal adding type enzyme. Indeed an enzyme capable of adding adenylate units to the 3'-ends of RNA chains has been isolated from the nuclei of a variety of mammalian cells [26,27]. It appears to be associated with 30S ribonucleoprotein particles and is RNA-dependent but not in the sense that it uses RNA as a 'template'. Rather RNA is used as a 'primer' for the sequential addition of adenylate units in an actinomycin D insensitive process using ATP as substrate (see Fig. 4.6). Whether this enzyme is actually responsible for polyA synthesis in mammalian cells has not yet been established. Nevertheless the *in vivo* process is apparently independent of DNA as a template, also being insensitive to actinomycin D. On the other hand, whilst this polyA synthesising activity was first detected

in nuclei, similar enzymes have been detected in cytoplasmic fractions. Perhaps such cytoplasmic activity is involved in the metabolic turnover of the 3'-polyA termini of mRNAs in the cytoplasm.

Another example of terminal addition enzyme activity is to be seen in the formation of the trinucleotide sequence pCpCpA common to all transfer RNAs (see Chapter 2). In the cell this sequence appears to have a metabolic turnover independent of the remainder of the transfer RNA molecule. The three nucleotides are apparently continuously removed and replaced [28].

Replacement involves the sequential addition of nucleotide units from their corresponding 5'-triphosphates catalysed by the action of separate cytoplasmic tRNA-nucleotidyl transferases operating as illustrated in Fig. 4.7. The cellular role of these nucleotide additions to tRNA is not understood, but it may serve in somewhay to regulate protein synthesis, perhaps by controlling the levels of functional tRNAs.

4.2.2 The sequence modifying enzymes

The enzymes catalysing the transfer of an intact methyl group (CH_3-) from S-adenosyl-L-methionine to a C, N or O atom of a purine or pyrimidine base or of ribose are called methylases [29]. Those specific for nucleotides in tRNA are the tRNA methylases, and those from

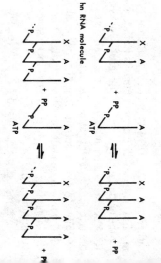

Fig. 4.6 The sequential addition of terminal units to hnRNA to generate polyA tracts.

Ribosomal RNA methylases, which act at the polynucleotide level, using S-adenosyl-L-methionine in the same way as the tRNA methylases, can be detected in *E. coli* as component parts of nascent ribosomal particles [32]. In mammalian cells the ribosomal RNA methylases appear to be associated with nucleoli, possibly in some association with the nascent ribosomal particles [33] by analogy with the *E. coli* situation.

Another type of enzymic modification is the formation of 4-thiouridine, another constituent of some tRNAs. A tRNA sulphur transferase has been isolated from *E. coli* which catalyses the *in vitro* transfer of sulphur from L-cysteine into tRNA uracil. ATP and Mg^{2+} are required [34]. (Δ^2-isopentenyl) adenosine on the other hand is formed in tRNA by the action of Δ^2-isopentenyl pyrophosphate tRNA-transferase. No cofactors are required and the enzyme (55 000 daltons) requires free-SH groups and is quite specific for tRNA [34]. Pseudouridine formation both in tRNA and ribosomal RNA, probably occurs by the enzymic modification of uridine residues in the RNA chain. Enzymes capable of carrying out this conversion have been detected in *E. coli* [34], and in the cytoplasm of mammalian cells [34]. The exact mechanism is not known. Whether it involves the cleavage of a N-glycoside bond in uridine, rotation of uracil residue, and the formation of a C-glycosidic linkage as shown in Fig. 4.8, remains to be seen.

4.2.3 *Cleaving and trimming enzymes*

Collectively such enzymes are in effect ribonucleases (RNAases) as they catalyse the hydrolysis of phosphodiester bonds in RNA molecules. Hydrolysis may be catalysed at points within the polymer, and if so, the RNAase is classed as an endonuclease. An exonuclease catalyses the stepwise hydrolysis of nucleotides from the end of a chain, either working from the 5'-end, towards the 3'-end,

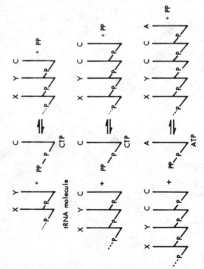

Fig. 4.7 Addition of terminal units to tRNA to generate sequence – pXpYpCpCpA.

E. coli can be considered in four groups. The uracil tRNA methylase and the cytosine tRNA methylase alkylate the C-5 position of the pyrimidine ring yielding ribothymine and 5-methylcytosine respectively. The adenine tRNA methylases give rise to 2-methyladenosine, 6-methyladenosine, and 6-dimethyladenosine. Two guanine tRNA methylases form 1-methylguanosine whilst a third enzyme is specific for the production of 7-methylguanosine [30].

S-adenosyl-   S-adenosyl-
methionine   homocysteine

  +   +

nucleotide   methylated
(in RNA)   nucleotide (in RNA)

$$\text{nucleotide (in RNA)} \xrightarrow{\text{methylase}} \text{methylated nucleotide (in RNA)}$$

tRNA methylase activity is also found in eukaryotes, predominantly in the cytoplasm [31]. Normally tRNA methylases are assayed by using methyl-deficient homologous tRNAs as substrates or normal tRNAs from heterologous sources which they can methylate to variable extents. Several neoplastic tissues are known to contain elevated levels of tRNA methylase activity and some virus or bacteriophage infections are known to affect tRNA methylase activities [29]. However the biological significance of these effects remains obscure.

or vice versa.

In *E.coli* two recently well characterised endonucleases are RNAase P and RNAase III [35]. They have similar ion requirements. RNAase P is found in association with ribosomes and is responsible for the removal of the extra 41 nucleotides from the tRNA precursor molecule illustrated in Fig. 3.8. A similar activity probably exists in mammalian cell cytoplasm [36] and can convert mammalian tRNA precursors to tRNA dimensions. RNAase III was found on the other hand to have a specific requirement for double helical RNA substrates. A role for this enzyme in the processing of large RNA transcripts was indicated from a use of *E.coli* mutants deficient in RNAase III [37]. In such mutants there is an accumulation of the 30S ribosomal precursor RNA. However if this primary transcript is isolated and treated with purified RNAase III, the two products 25S and 17.5S result (see Chapter 3, Fig. 3). In addition to a role in ribosomal RNA processing it is probably involved in mRNA production in T7 bacteriophage infected *E.coli* [38]. As mentioned in Chapter 3, the five early T7 mRNAs are transcribed as a large unit. This

large unit accumulates in the RNAase III deficient *E.coli* mutant but if isolated and treated subsequently in the test-tube with purified RNAase III the five early mRNAs are produced (see Fig. 3.5).

Another important *E.coli* enzyme is RNAase II [35] which is an exonuclease working in the 3'→5' direction. The 5'-terminal fragment of the tyrosine tRNA precursor produced by RNAase cleavage is probably degraded to mononucleotides by this enzyme. Also the two extra nucleotides at the 3'-end of the precursor are probably removed by this RNAase II. Another role proposed for RNAase II is the trimming of the 25S and 17.5S *E.coli* intermediate ribosomal precursors to the final 23S and 16S dimensions (see Chapter 3, Fig. 3).

The situation with regard to similar RNAases in higher organisms is less clear cut. A RNAase III-like activity has been detected in HeLa cell nuclei [39] but whether it can be shown to be involved in 45S RNA or hnRNA processing remains to be seen. For the main part, most *in vitro* studies aimed at elucidating the enzymic cleavage of the 45S precursor have so far been confined to studying events as they take place

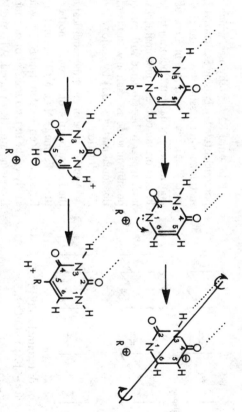

Fig. 4.8 A possible mechanism for the formation of pseudouridine.

in the 80S nascent ribosomal precursor particles as isolated from nucleoli [40]. Addition of purified 45S RNA to isolated nucleolar preparations has resulted only in the complete non-specific breakdown of the substrate RNA.

References

[1] Hurwitz, J. and August, J.T. (1963), *Prog. Nucleic Acid Res. Mol. Biol*, **1**, 59.

[2] Geiduschek, E.P., Nakamoto, T. and Weiss, S.B. (1961), *Proc. Nat. Acad. Sci. U.S*, **47**, 1405.

[3] Burgess, R.R. (1971), *Ann. Rev. Biochem*, **40**, 711.

[4] Travers, A. (1974), *Cell*, **3**, 97.

[5] Dickson, R.C., Abelson, J., Barnes, W.M. and Reznikoff, W.S. (1975), *Science*, **187**, 27.

[6] Saucier, J.M. and Wang, J.C. (1972), *Nature New Biol*, **239**, 167.

[7] Bautz, E.K.F. (1972), *Prog. Nucleic Acid Res. Mol. Biol*, **12**, 129.

[8] Heil, H. and Zillig, W. (1970), *FEBS Letts*, **11**, 165.

[9] Goldthwait, D.A., Antony, D.D. and Wu, C-W. (1970), in *Lepetit Colloquium on RNA polymerase and transcription*, Ed. L. Silvestri, p. 10, Wiley, New York.

[10] Maitra, U. and Hurwitz, J. (1965), *Proc. Nat. Acad. Sci. U.S*, **54**, 815.

[11] Gilbert, W. and Maxam, A. (1973), *Proc. Nat. Acad. Sci. U.S*, **70**, 3581.

[12] Goldberg, I.H. and Friedman, P.A. (1971), *Ann. Rev. Biochem*, **40**, 775.

[13] Roberts, J.W. (1969), *Nature*, **224**, 1168.

[14] Oda, T. and Takanami, M. (1972), *J. Mol. Biol*, **71**, 799.

[15] Weiss, S.B. (1950), *Proc. Nat. Acad. Sci. U.S*, **46**, 1020.

[16] Burdon, R.H. and Smellie, R.M.S. (1962), *Biochim. Biophys. Acta*, **61**, 633.

[17] Jacob, S.T. (1973), *Prog. Nucleic Acid Res. Mol. Biol*, **13**, 93.

[18] Roeder, R.G. and Rutter, W.J. (1970), *Proc. Nat. Acad. Sci. U.S*, **65**, 675.

[19] Price, R. and Penman, S. (1972), *J. Mol. Biol*, **70**, 435.

[20] Stein, H. and Hausen, P. (1971), *Cold Spring Harb. Symp. Quant. Biol*, **35**, 709.

[21] Weaver, R.F., Blatti, S.P. and Rutter, W. J. (1971), *Proc. Nat. Acad. Sci. U.S*, **68**, 2994.

[22] Sklar, V.E.F., Schwartz, L.B. and Roeder, R.G. (1975), *Proc. Nat. Acad. Sci. U.S*, **72**, 348.

[23] Kunztzel, H. and Schafer, K.P. (1971), *Nature*, **231**, 265.

[24] Scragg, A.H. (1971), *Biochem. Biophys. Res. Commun*, **45**, 701.

[25] Saccone, C., Gallerani, R., Gadieta, M.N. and Greco, M. (1971), *FEBS Letts*, **18**, 339.

[26] Edmonds, M. and Abrams, R. (1960), *J. Biol. Chem*, **235**, 1142.

[27] Burdon, R.H. (1963), *Biochem. Biophys. Res. Commun*, **11**, 472.

[28] Deutcher, M.P. (1973), *Prog. Nucleic Acid Res. Mol. Biol*, **13**, 51.

[29] Borek, E. and Srinivasan, P.R. (1966), *Ann. Rev. Biochem*, **35**, 275.

[30] Hurwitz, J., Gold, M. and Anders, M. (1964), *J. Biol. Chem*, **239**, 3462.

[31] Burdon, R.H., Martin, B.M. and Lal, B. (1967), *J. Mol. Biol*, **28**, 357.

[32] Thammana, P. and Held, W.A. (1974), *Nature*, **251**, 682.

[33] Culp, L.A. and Brown, G.M. (1970), *Arch. Biochem. Biophys*, **137**, 222.

[34] Soll, D. (1971), *Science*, **173**, 293.

[35] Altman, S. and Robertson, H.D. (1973), *Molec. Cell Biochem*, **1**, 83.

[36] Burdon, R.H. (1971), *Prog. Nucleic Acid. Res. Mol. Biol*, **11**, 33.

[37] Nickolaev, N., Schlessinger, D. and Wellauer, P.K. (1974), *J. Mol. Biol*, **86**, 741.

[38] Dunn, J.J. and Studier, F.W. (1973), *Prog. Nat. Acad. Sci. U.S*, **70**, 1559.

[39] Kwan, C.M., Gotoh, S., Schlessinger, D.

(1974), *Biochim. Biophys. Acta*, **349**, 428.

[40] Winicov, I. and Perry, R.P. (1974),

Biochemistry, **13**, 2908.

[41] Loewen, P.C., Sekiya, T. and Kohrana, H.G. (1974), *J. Biol. Chem.*, **249**, 217.

5 The control of RNA production

The purpose of this last chapter is to summarise quite briefly the wide range of molecular mechanisms whereby the production of various RNAs within the cell could be regulated. These, as will become obvious, can operate at all levels of the overall process.

5.1 The synthesis of RNA polymerase itself

A bacterial culture growing in the steady state synthesises RNA in precise coordination with other macromolecular species [1]. Recent analysis of the RNA synthesised by E.coli in different growth states suggests that a cell may effect this coordination by regulating the rate of RNA polymerase synthesis [2]. If it were not regulated the rate of synthesis of the individual subunits would simply double as their genes were replicated. In the case of the β and β' subunits of E.coli this does not appear to be the case. The genes for these subunits appear to be contained together in an independently regulated operon, and their products are synthesised sequentially and coordinately. How the regulation of this operon is actually coordinated with cell growth is however not yet known [3].

5.2 Factors involved in RNA chain initiation

Normally, nucleotide units are added to a growing RNA chain at a rate of 40 to 50 nucleotides per second at 37° (in E.coli). This rate, however, can vary with temperature and with other environmental changes, but under normal conditions the amount of RNA made

in a bacterium is limited, not so much by the rate of growth of RNA chains, but by their rate of initiation. This varies quite considerably for individual types of RNA molecule. Ribosomal RNA molecules for instance, are required in fairly large amounts and can be initiated in E.coli at the rate of one molecule every second. On the other hand a gene coding for a protein present in very small amounts may be transcribed as infrequently as once every bacterial generation. Having initiated an RNA chain, the RNA polymerase moves away from the promoter site transcribing the adjacent genetic material and leaves the initiation site open to a second polymerase molecule. The frequency of these initiations will determine the proximity of RNA polymerase molecules on the genomic sequences in question. In the case of the ribosomal genes this may be as close together as is sterically possible.

How is the frequency of initiation regulated? Considerable insight into the problem has come from the study of the factors controlling the synthesis of the mRNAs for the enzyme β-galactosidase and other enzymes coded for by the lac operon in E.coli [4] (see R.A. Woods' Biochemical Genetics in this series). Transcription of this operon for instance can be controlled negatively and positively [5]. Negative control is mediated by the lac repressor which binds specifically and tightly to the operator region (o), thereby preventing transcription. Positive control of the lac operon is exerted through the phenomenon termed catabolite repression. Expression of the *lac* operon (and other catabolite repressible operons) is repressed when glucose (a more efficient source of carbon than lactose) is present in the medium. By a yet unknown mechanism, the presence of glucose results in a decreased concentration of intracellular cyclic AMP. Cyclic AMP is required for efficient expression of the *lac* operon, since it activates the catabolite gene activator protein (CAP), which in turn activates initiation of transcription of *lac* by RNA polymerase [6]. This CAP seems to interact with a site in the promoter region (p) of the operon (see Figs. 2 and 5.1).

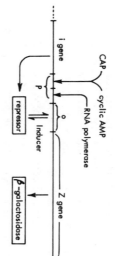

Fig. 5.1 The *lac* region of the *E.coli* chromosome.

CAP can be envisaged as a dimer of two special subunits (mol. wt.40 000) each equivalent to 11 nucleotide pairs in diameter. The CAP binding site is located in the promoter region of two-fold symmetry noted in Fig. 4.2 of the previous chapter [5]. The binding of this protein-cyclic AMP complex may then destabilise the guanine plus cytosine rich region next to it. This would lower the transition temperature of the entry site for the RNA polymerase some 14 nucleotide pairs away (the entry site is positioned at the adenine plus thymine rich site shown in Fig. 4.2). RNA polymerase entry is thereby facilitated. Additional nucleotide sequence analyses [7,8] have been recently carried out on the operator region (o) lying between the promoter region and the start of the z gene coding for β-galactosidase. Once again a region of two fold symmetry will be noted (Fig. 5.2)

[5]. The significance of this is not yet clear but the *lac* repressor appears to act as a tetramer [9]. Kinetic studies on repressor-DNA and repressor-inducer complex formation suggest that inducers act by dissociating the repressor-operator complex, rather than by causing direct transconformation to free repressor molecules [9].

In the absence of repressor binding to the operator (e.g. in the presence of inducer), the question arises whether the operator sequence is transcribed? As a result of studies on the *lac* operon mRNA sequences [8] two interesting features emerge, (a) a portion, but not all, of the operator sequence seems to be transcribed, and (b) the actual 'start' site for RNA synthesis is some 35 nucleotide pairs away from the 'entry' site in the promoter. Thus there must be some 'drift' of the polymerase to the 'start' site after entry (see Fig. 5.3). Recent interesting data indicate the first 33 nucleotides of both the *lac* mRNA and the *gal* mRNA to be identical [28].

Initiation can be effected in other ways. For example ribosomal RNA in bacteria such as *E.coli* in the rapid growth state can comprise up to 50 per cent of the RNA being made. However, when these bacteria have exhausted the amino acid content of the medium, the proportion of rRNA synthesis can be ten fold lower.

Two unusual nucleotides, guanosine tetraphosphate (ppGpp) and guanosine pentaphosphate (pppGpp) may serve in some way to regulate the level of rRNA biosynthesis [11]. The intracellular concentration of ppGpp appears to be inversely correlated with the rate of rRNA synthesis, and its synthesis results from an ATP-dependent conversion of GDP. This may represent an idling translational reaction occurring on the messenger-ribosome complexes, which could be 'signalled' by the presence of uncharged tRNA [9]. How ppGpp effects rRNA initiation is presently a matter of controversy. However, Travers and his col-

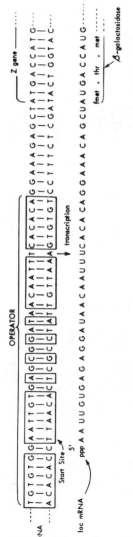

Fig. 5.2 The *lac* operator region and the z gene showing the nature of the 5'-end of the *lac* mRNA.

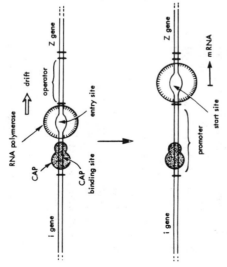

Fig. 5.3 A possible mode of initiation of *lac* transcription.

leagues [10] report the existence of psi (ψ) factor, a possible controlling element specifically acting on the initiation of ribosomal RNA during the *in vitro* transcription of *E.coli* DNA. In the presence of KCl this factor stimulates the rate of rRNA synthesis up to ten fold. However, the ψ-dependent stimulation effects could be abolished by ppGpp. Recent reports suggest that ψ is equivalent to the Tu-Ts elongation factor in protein synthesis (see A.E. Smith's *Protein Biosynthesis* in this series). Thus there is the possibility of a link between translation and rRNA synthesis.

5.3 Structural modifications to RNA polymerases

During infection of *E.coli* with T4 bacteriophage, host RNA synthesis rapidly stops and a well defined series of changes in the pattern of T4 DNA transcription (e.g. 'early' to 'late' RNA) occurs, which may result from changes observed in the polypeptides that make up the RNA polymerase subunit structure. Within 4 min of infection, the two α-subunits are enzymically modified. The modification involves the covalent attachment of an adenosine diphosphoribose unit to each subunit. The attachment appears to be through its terminal ribose to a guanido nitrogen of a specific arginine residue [12], the donor being NAD$^+$.

Another type of structural modification to the *E.coli* polymerase that occurs during T4 infection is the binding of several bacteriophage specific proteins, the products of T4 genes 33 and 55. (The product of gene 55 is a 22 000 molecular weight binding protein [13]). Their actual mode of action is not yet understood, however it is probable that auxiliary transcription factors regulate the specificity of bacterial RNA polymerases. Indeed it has been observed that free RNA polymerase in crude extracts of uninfected bacteria is both physically and functionally heterogeneous [14], two or perhaps three forms of the enzyme being distinguishable.

In the case of spore formation by *B. subtilis* there is a change in the pattern of transcription.

53

A comparison of the RNA polymerase from vegetative cells and from sporulating cells reveals some structural differences [15]. The polymerase from the sporulating cells does not have a tightly bound σ subunit, but appears to have associated with it additional small polypeptides. Thus the change in transcription pattern during sporulation is probably a consequence of structural modification of the enzyme.

5.4 The complexity of RNA polymerases

When the bacteriophage T7 infects its host *E.coli*, the RNA polymerase of the host transcribes a portion of the invading genome corresponding to the 'early' genes (see Fig. 3.11 in Chapter 3). The product of one of these genes is an mRNA which, when translated by the host's protein synthetic machinery, yields a very simple RNA polymerase (a single polypeptide chain of molecular weight about 110 000) [16]. This T7 specified RNA polymerase can only initiate RNA synthesis at promoter sites on T7 DNA other than those at which the host's RNA polymerase can initiate. In this way the T7 specified polymerase, with a different initiation specificity, simply synthesises a collection of RNAs known as the 'late' mRNAs, from the remainder of the T7 genome. The reason for the apparent structural simplicity of the T7 polymerase is not understood. It may be, that in some way, complexity of polymerase structure reflects the organisms requirement for higher degrees of control. The host, *E.coli*, is capable of rapid adaptation to different growth conditions, and perhaps this is reflected in the greater complexity of its polymerase, the initiation specificity of which can be altered by interaction with regulatory proteins. A similar situation is known to prevail in *B. subtilis* which can alter its growth pattern quite dramatically (e.g. spore formation as already mentioned). On the other hand, some bacteria like the *halophilic* group have a fairly stable growth pattern and appear to have simple type polymerases. From Chapter 4 it will be recalled that the RNA polymerases of eukaryotic cell nuclei belong to the complex type, whilst the polymerases from the mitochondria are simple single polypeptides. Perhaps there is little need for extensive regulation of mitochondrial DNA transcription. Indeed it appears that both strands are transcribed (see Chapter 3) and extensive control may operate instead at the post-transcriptional level.

5.5 The variety of RNA polymerases

A striking difference between the transcriptional equipment of eukaryotic cells and prokaryotic cells, is the presence in the latter of more than one variety of polymerase. From Chapter 4 it will be remembered that so far there have been detected three distinguishable varieties of complex type polymerases in some mammalian cell nuclei. Even one of the simplest of the nucleated organisms, yeast, has at least two complex nuclear polymerases. It would appear in these cases, that the initiation specificity (i.e. for rRNA, tRNA, hnRNA etc.) may lie not only in the complexity of the polymerases, but also in their variety and/or intranuclear location. A question being pursued at present is whether the levels of the various polymerases can regulate the levels of the various RNAs produced. In higher eukaryotes qualitative changes in RNA synthesis can be induced by hormone treatments, nutritional deficiencies etc. The levels of activity of the various purified polymerases have been shown to be different, for instance, when extracted from the hormone-treated organisms. These alterations appear to result not from increases in the *de novo* synthesis of RNA polymerases, but rather from the activation of preexisting molecules [31] (e.g. possibly by interaction with hormone-receptor complexes).

5.6 Gene dosage

Another means of increasing the number of transcripts of a particular region of the genome is to have the sequences in question represented a number of times within the genome. This, as already mentioned in Chapter 3, is the case for rRNA, tRNA and 5S genes. The actual numbers of copies of these genes seems to increase with the organism's evolutionary complexity. Most mRNAs on the other hand arise from sequences that occur once (or only a few times) in the genome. Exceptions are the genes for histone mRNA which occur in multiple copies [17], as do those for chick feather keratin mRNAs [29].

Of particular interest has been the question of whether the large number of say, rRNA genes, is constant throughout the life-cycle of a cell; or whether they are amplified *de novo* by a mechanism sensitive to metabolic demands. One case where amplification of the ribosomal genes has been demonstrated is in the immature oocytes of amphibia and insects [18]. rRNA synthesis reaches a peak rate in such immature oocytes which are in the pachytene and di-plotene stages of meiosis, and it is during these stages that extra copies of the nuclear organiser region are synthesised selectively to give rise to about 10^6 copies of the ribosomal genes (somatic *Xenopus* cells have about 1600 copies).

5.7 Chromosomal organisation

Whilst both the prokaryotes and eukaryotes have complex polymerases, the physical organisation of their chromosomes differs. In those prokaryotes which have been studied in detail the chromosomes can best be considered as helical DNA molecules (cyclic in the case of *E.coli*) to which can be associated at various sites such proteins as repressors, cyclic AMP proteins, etc. These, as already indicated, control the nature of the transcripts produced. Eukaryotic chromosomes on the other hand are more complex. They are almost completely associated with various proteins, and constitute a complicated deoxyribonucleoprotein complex

referred to, in interphase nuclei, as *chromatin*. There is general agreement that a single DNA double helix (20 Å) combines with protein to give a 30 Å fibre which is then variously aggregated and condensed to give thicker, tightly packaged structures. This aggregation is proposed to occur with local condensations resulting in the formation of 80 Å–100 Å diameter structures giving fibres, when slightly extended, a 'string of beads' structure [19]. These beads are basic repeat units comprising 200 nucleotide stretches of DNA combined with basic histone molecules (two each of types F2A1, F3, F2A2 and F2B) [20], possibly as shown in Fig. 5.4.

Studies aimed at elucidating the functional significance of the proteins associated with the DNA have relied on *in vitro* transcription of chromatin with added bacterial RNA polymerases. Briefly summarised, these experiments indicate that certain genomic sequences are in some way 'masked' in the chromatin complex [21]. The means whereby this 'masking' is achieved is not known precisely, but it is tempting to equate the tightly packaged structures with functionally inactive DNA and the more open chromatin fibres with active DNA. However, neither type of region by itself contains enough DNA to accommodate even a single gene. An attractive proposal is that all DNA is bound up in 'bead' structures except where it is necessary to have open chromatin. The DNA in these open regions might represent 'control' DNA which is accessible to the outside environment (e.g. hormones etc.), and which could regulate the activity of the genes with which it is associated. For instance, in response to the appropriate signal, the 'control' regions might permit the RNA polymerase(s) to begin transcribing the gene itself which is in the tightly packed bead-like structures. This would then open up and also become available for copying. Most biochemical data indicate that histones are certainly involved in repres-

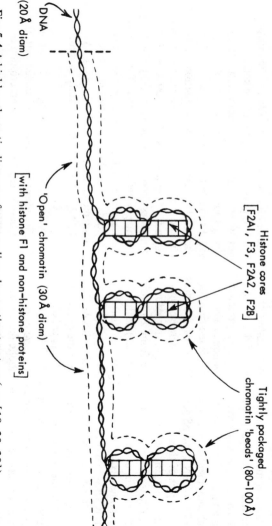

Fig. 5.4 A highly schematic diagram of mammalian chromatin structure (see [19, 20, 30]).

DNA (20 Å diam)

'Open' chromatin (30 Å diam) [with histone F1 and non-histone proteins]

Histone cores [F2AI, F3, F2A2, F2B]

Tightly packaged chromatin 'beads' (80-100 Å)

sing the availability of certain sequences of eukaryotic DNA in chromatin for transcription. Control of gene transcription appears to be a positive event, involving the de-repression of histone action by the non-histone protein components also associated with chromatin [22]. From Fig. 5.4 no particular location has been ascribed to the non-histone proteins of chromatin, nor indeed for the remaining histone F1. At present it seems that both F1 and non-histone proteins are associated with the 'open' regions but some non-histones may be associated with the 'beads'.

Readers wishing a fuller account of chromatin structure and its role in cell differentiation are referred to Cell Differentiation by J.M. Ashworth, in this series. Additionally there are many recent data implicating non-histone chromatin proteins as mediators in a number of specific hormone induced gene activations. A more detailed coverage of this can be found in Hormone Action, by A. Malkinson, also in this series.

The DNA unmasking effects envisaged above could be regarded as a 'coarse' type of control.

In addition there may well be a number of 'fine' levels of control, as yet undiscovered in eukaryotes, for example, regulating correct initiation, termination etc.

5.8 Post-transcriptional processing

The variety of post-transcriptional processes operating in the production of various RNAs of eukaryotic and prokaryotic origin clearly offers considerable potential vis-à-vis the further regulation of mature, functional RNA production. To date, however, very little is known about the control of these processing events. Cleavage rates of the ribosomal RNA precursors in mammalian cells can be altered by hormone administration, virus infections, and may also be linked to ribosomal protein synthesis [23]. Cleavage of tRNA precursors in vivo can be influenced by cell growth conditions and infections by some viruses [23]. With regard to mRNA production, there is evidence indicating increased rates of hnRNA processing during the transition of mammalian cells from the 'resting' to the 'growing' state [24]. Despite this information it is nevertheless unclear precisely what regulates the cleavage enzymes. On the other hand, the modification of tRNAs has been shown to vary in different biological situations, and naturally occurring inhibitors of the methylases involved have been detected in some eukaryotes [25]. Post-transcriptional addition of nucleotides can also be modulated in certain situations.

For instance the activation of protein synthesis after sea urchin fertilization, is accompanied by polyadenylation of the stored mRNAs of the egg [26]. The effect is to increase the tracts of adenylate residues from about 100 to 200. How this is regulated, and indeed what its significance may be is not yet appreciated.

A final post-transcriptional process that must not be forgotten, but which will have a profound effect on, say, the levels of cellular mRNAs, is of course their metabolic degradation. This is an area for further study, but there is evidence that degradation of mRNA takes place in a ribosomal complex, and is initiated on parts of the messenger just translated. A possible mechanism involves recognition of regions of special structure by endonucleases, followed by exonuclease digestion [27].

5.9 Concluding remarks

Given the complexity and multiplicity of reactions involved in the biosynthesis and maintenance of functional cellular RNA molecules, it is perhaps not surprising to have encountered such a diverse set of potential regulatory mechanisms. Their importance relative to one another may depend on the particular biological context, but as a generalisation it now seems clear that a change in the intracellular level of a particular RNA will almost always involve several regulatory mechanisms operating in concert.

References

[1] Maaløe, O. and Kjeldgaard, No.O. (1966) Control of Macromolecular Synthesis: a study of DNA, RNA and protein synthesis in bacteria, Benjamin, New York.

[2] Nierlich, D.P. (1972), J. Mol. Biol., 72. 751.

[3] Scaife, J.G. (1973), Brit. Med. Bull., 29, 214.

[4] Jacobs, F. and Monod, J. (1961), J. Mol. Biol., 3, 318.

[5] Dickson, R.C., Abelson, J., Barnes, W.M. and Reznikoff, W.S. (1975), Science, 187, 27.

[6] de Crombrugghe, B., Chen, B., Anderson, W., Nisley, P., Gottesman, M., Pastan, I. and Perlman, R. (1971), Nature New Biol., 231, 139.

[7] Gilbert, W. and Maxam, A. (1973), Proc. Nat. Acad. Sci. U.S., 70, 3581.

[8] Maizels, N. (1973), Proc. Nat. Acad. Sci. U.S., 70, 3581.

[9] Gros, F. (1974), FEBS Letts., 40, 519.

[10] Travers, A., Kamen, D. and Cashel, M. (1970), Cold Spring Harb. Symp. Quant. Biol., 35, 415.

[11] Cashel, M. (1969), J. Biol. Chem., 244, 3133.

[12] Goff, C.G. (1974), J. Biol. Chem., 249, 6181.

[13] Ratner, D. (1974), J. Mol. Biol., 89, 803.

[14] Travers, A. and Buckland, R. (1973), Nature New Biol., 243, 251.

[15] Greenleaf, A., Linn, T. and Losick, R. (1973), Proc. Nat. Acad. Sci. U.S., 70, 490.

[16] Chamberlin, M., McGrath, J. and Waskell, L. (1970), Nature, 228, 227.

[17] Kedes, L.H., Birnstiel, M.L. (1971), Nature New Biol., 230, 165.

[18] Birnstiel, M.L., Chipchase, M. and Spiers, J. (1971), Prog. Nucleic Acid Res. Mol. Biol., 11, 351.

[19] Olins, A.L. and Olins, D.E. (1974), Science 183, 330.

[20] Kornberg, R. (1974), Science, 184, 865 and 868.

[21] Gilmour, R.S. and Paul, J. (1969), J. Mol. Biol., 34, 305.

[22] Stein, G.S., Stein, J.S. and Kleinsmith, L.J. (1975), *Scientific American*, **232**, 46.

[23] Burdon, R.H. (1971), *Prog. Nucleic Acid Res. Mol. Biol.*, **11**, 33.

[24] Johnson, L.F., Abelson, H.T., Green, H. and Penman, S. (1974), *Cell*, **1**, 95.

[25] Kerr, S.J. (1971), *Proc. Nat. Acad. Sci. U.S.*, **68**, 406.

[26] Slater, D.W., Slater, I. and Gillespie, D. (1972), *Nature*, **240**, 333.

[27] Altman, S. and Robertson, H.D. (1973), *Mol. Cell. Biochem.*, **1**, 83.

[28] Musso, R.E., de Crombrugghe, B., Pastan, I., Sklar, J., Yot, P. and Weissman, S. (1974), *Proc. Nat. Acad. Sci. U.S.*, **71**, 4940.

[29] Kemp, D.J. (1975), *Nature*, **254**, 573.

[30] Van Holde, K.E., Sahasrabuddhe, E.G. and Shaw, R.B. (1974), *Nucleic Acids Res.*, **1**, 1579.

[31] Jacob, S.T. (1973), *Prog. Nucleic Acid Res. Mol. Biol.*, **13**, 93.

Addenda

Polarity of the transcription product of eukaryotic ribosomal genes

The illustrations of the eukaryotic ribosomal RNA precursors shown in Fig. 3.2 (HeLa) and Fig. 3.3 (*Xenopus*) would imply the sequences corresponding to the 28S ribosomal RNA to be at the 5'-ends of the precursor molecules (i.e. left hand side of diagrams). There is however some controversy concerning this point. Recent data from *in vitro* transcription studies, drug and UV sensitivity studies (see R.P. Perry, (1976) *Ann. Rev. Biochem.* **45**, in press) suggest the sequence polarity of the precursors to be the reverse of what is indicated. In other words the order of sequences in the precursors is now believed to be 5' *end*-spacer-18S sequence-spacer-5.8S sequence-spacer-28S sequence-3' *end*. This order is more in line with the prokaryotic situation where the sequence corresponding to the 16S component is transcribed before the 23S component (see Fig. 3.5) and thus is towards the 5'-end of the prkaryotic precursor.

Histone nomenclature

In Fig. 5.4 and on pages 55 and 56 various histone classes are referred to as F1, F2A1, F2A2, F2B and F3. To accord with recent changes in nomenclature these types should now be referred to as H1, H2A, H2B, H3 and H4 respectively.

Index

574.8732 Burdon, Roy
B   Hunter.
     RNA biosynthesis

92794

574.8732 Burdon, Roy
B   Hunter.
     RNA biosynthesis 92794

DATE	BORROWER'S NAME
9/2/81	Kathryn Weeks

92794